BARODESY AND ITS APPLICATION FOR CLAY

T0135552

ADVANCES IN GEOTECHNICAL ENGINEERING AND
TUNNELLING

20

General editor:

D. Kolymbas

University of Innsbruck, Division of Geotechnical and Tunnel Engineering

In the same series (A.A.BALKEMA):

1. D. Kolymbas (2000), *Introduction to hypoplasticity*, 104 pages, ISBN 90-5809-306-9

2. W. Fellin (2000), *Rütteldruckverdichtung als plastodynamisches Problem (Deep vibration compaction as a plastodynamic problem)*, 344 pages, ISBN 90-5809-315-8

3. D. Kolymbas & W. Fellin (2000), *Compaction of soils, granulates and powders - International workshop on compaction of soils, granulates, powders,* Innsbruck, 28-29 February 2000, 344 pages, ISBN 90-5809-318-2

In the same series (LOGOS):

4. C. Bliem (2001), *3D Finite Element Berechnungen im Tunnelbau (3D finite element calculations in tunnelling)*, 220 pages, ISBN 3-89722-750-9

5. D. Kolymbas, ed. (2001), *Tunnelling Mechanics, Eurosummerschool, Innsbruck, 2001*, 403 pages, ISBN 3-89722-873-4

6. M. Fiedler (2001), *Nichtlineare Berechnung von Plattenfundamenten (Nonlinear Analysis of Mat Foundations)*, 163 pages, ISBN 3-8325-0031-6

7. W. Fellin (2003), *Geotechnik - Lernen mit Beispielen*, 230 pages, ISBN 3-8325- 0147-9

8. D. Kolymbas, ed. (2003), *Rational Tunnelling, Summerschool, Innsbruck 2003*, 428 pages, ISBN 3-8325-0350-1

9. D. Kolymbas, ed. (2004), *Fractals in Geotechnical Engineering, Exploratory Workshop, Innsbruck, 2003*, 174 pages, ISBN 3-8325-0583-0

10. P. Tanseng (2006), *Implementation of Hypoplasticity for Fast Lagrangian Simulations*, 125 pages, ISBN 3-8325-1073-7.

11. A. Laudahn (2006), *An Approach to 1g Modelling in Geotechnical Engineering with Soiltron*, 197 pages, ISBN 3-8325-1072-9.

12. L. Prinz von Baden (2005), *Alpine Bauweisen und Gefahrenmanagement*, 212 pages, ISBN 3-8325-0935-6.

13. D. Kolymbas, A. Laudahn, eds. (2005), *Rational Tunnelling, 2nd Summerschool, Innsbruck 2005*, 291 pages, ISBN 3-8325-1012-5.

14. T. Weifner (2006), *Review and Extensions of Hypoplastic Equations*, 240 pages, ISBN 978-3-8325-1404-4.

15. M. Mähr (2006), *Ground movements induced by shield tunnelling in non-cohesive soils*, 168 pages, ISBN 978-3-8325-1361-0.

16. A. Kirsch (2009), *On the face stability of shallow tunnels in sand*, 178 pages, ISBN 978-3-8325-2149-3.

17. D. Renk (2011), *Zur Statik der Bodenbewehrung*, 165 pages, ISBN 978-3-8325-2947-5.

18. B. Schneider-Muntau (2013), *Zur Modellierung von Kriechhängen*, 225 pages, ISBN 978-3-8325-3474-5.

19. A. Blioumi (2014), *On Linear-Elastic, Cross-Anisotropic Rock*, 215 pages, ISBN 978-3-8325-3584-1.

Barodesy and its Application for Clay

Gertraud Medicus
University of Innsbruck, Division of Geotechnical and Tunnel Engineering

E-mail: Gertraud.Medicus@uibk.ac.at
Homepage: http://www.uibk.ac.at/geotechnik

The first three volumes have been published by Balkema
and can be ordered from:

A.A. Balkema Publishers
P.O.Box 1675
NL-3000 BR Rotterdam
e-mail: orders@swets.nl
website: www.balkema.nl

This publication has been produced with financial support of
the Vice Rector for Research of the University of Innsbruck.

Bibliographic information published by the Deutsche Nationalbibliothek

The Deutsche Nationalbibliothek lists this publication in the Deutsche
Nationalbibliografie; detailed bibliographic data are available in the
Internet at http://dnb.d-nb.de.

ISBN 978-3-8325-4055-5

ISSN 1566-6182

Logos Verlag Berlin GmbH
Comeniushof, Gubener Str. 47,
10243 Berlin
Tel.: +49 030 42 85 10 90
Fax: +49 030 42 85 10 92
INTERNET: http://www.logos-verlag.de

Acknowledgement

I am grateful to my supervisor Prof. Dimitrios Kolymbas for supporting me over the last years. He was always interested in my research work, was available for discussion about constitutive modelling every day and provided valuable assistance.

I thank Prof. Ivo Herle for his important and useful feedback. Every single comment improved the value of this work.

I also thank

- Wolfgang Fellin for being an appreciated adviser and for his contribution in many important discussions,

- Barbara Schneider-Muntau for her keen perception and for countless productive discussions on soil mechanics and tensor calculus,

- Ansgar Kirsch for introducing me in the topic of constitutive modelling,

- Chien-Hsun Chen for providing (numerical) support over the last years,

- Stefan Rainer (*Workgroup Numerical Analysis, Department of Mathematics*) who was available for mathematical advise,

- Kornelia Nitzsche and Vladislava Kostkanová from *TU Dresden* for providing laboratory test results,

- Sarah-Jane Loretz-Theodorine and Christine Neuwirt for their administrative support,

- Daniel Renk, Anastasia Blioumi, Iliana Polymerou, Stefan Tilg, Franz Haas, Franz Berger, Iman Bathaeian and Josef Wopfner for the pleasant working atmosphere and their generous readiness to help others.

In the last months, I received financial support by a doctoral scholarship (*Doktoratsstipendium aus der Nachwuchsförderung*) from the *University of Innsbruck*. My research activity was further supported by the funding *Hypo Tirol Bank Forschungsförderungspreis*. This publication has been produced with financial support of the *Vice Rector for Research of the University of Innsbruck*. I thank the *University of Innsbruck* and *Hypo Tirol Bank* for granting financial support.

Thank you, Daniel, for the invaluable personal support!

Kurzfassung

Die *Barodesie* ist ein neues Materialmodell zur Beschreibung von Bodenverhalten und kann als Weiterentwicklung der Hypoplastizität gesehen werden. Der mathematischen Formulierung liegt das asymptotische Bodenverhalten zugrunde: Wird Boden durch einen *proportionalen Deformationspfad* verformt, resultiert daraus ein Spannungspfad, der sich asymptotisch einem *proportionalen Spannungspfad* nähert.

Im Zuge dieser Arbeit wurde ein vorhandener Ansatz, der proportionale Deformationspfade mit proportionalen Spannungspfaden verknüpft, erweitert. Die Ergebnisse werden einigen weit verbreiteten *Spannungs-Dilatanz Beziehungen* gegenübergestellt.

In weiterer Folge wurde die *Barodesie für Ton* entwickelt. Bodenmechanische Eigenschaften, wie kritische Zustände oder der Einfluss des Spannungsniveaus und der Lagerungsdichte, sind im Modell enthalten. Die Kalibrierung ist einfach: Fünf bekannte Bodenparameter werden benötigt und können anhand eines konsolidierten, undrainierten Triaxialversuchs bestimmt werden. In der Arbeit werden Elementversuche numerisch simuliert und Laborversuchsergebnissen von verschiedenen Tonarten gegenübergestellt. Außerdem dienen zwei Materialmodelle, das *hypoplastische Modell von Mašín* und das elasto-plastische *Modified Cam Clay model*, als Vergleich.

Abstract

Barodesy is a new constitutive frame for soil and can be seen as a further development of Hypoplasticity. The basis of the mathematical formulation is given by the asymptotic behaviour of granulars. If soil is deformed with a *proportional strain path*, the resulting stress path approaches asymptotically a *proportional stress path*.

In the course of this thesis an existing relation which links proportional strain paths to proportional stress paths was further developed. The results are compared with several *stress-dilatancy relations*.

A modification of *barodesy* to model clay behaviour is introduced. Common concepts of soil mechanics, such as critical states, barotropy and pyknotropy are comprised. The calibration of *barodesy for clay* is simple, five well known soil parameters, which are obtained by a consolidated undrained triaxial compression test, are sufficient.

Standard element tests are simulated with *barodesy* and compared with experimental data of different clay types. The numerical simulations are also compared with two frequently used constitutive models for clay, the *hypoplastic model for clay by Mašín* and the elasto-plastic *Modified Cam Clay model*.

Contents

Chapter 1

Introduction

A constitutive model is a theory which describes (with some simplifications) stress-strain relations of a material and which should enable an acceptable representation of reality. It is difficult to maintain an overview of the great number of constitutive models for soils proposed so far. Soil is often modelled as an elastic and/or perfectly plastic material. In order to capture variable strength and stiffness, hardening plasticity models have been developed where density and preloading effects are taken into account. Some bounding surface plasticity models are reported to provide realistic simulations of cyclic behaviour, e.g [29, 60].

The development of *hypoplastic models* started in the 1970s. They do not use the standard notions of elasto-plastic models (such as elastic region, yield function, flow function, ...). Instead, the effective stress rate $\overset{\circ}{\mathbf{T}}$ is given as a function of stretching $\mathbf{D}$ and the effective stress $\mathbf{T}$, optionally complemented by internal variables[1]. In other words, the mathematical formulation is a differential equation: $\overset{\circ}{\mathbf{T}} = \mathbf{f}(\mathbf{T}, \mathbf{D}, ...)$.

1.1   Developments in the field of hypoplastic models

Hypoplastic equations are characterized by the fact that incremental soil response can be described by a single rate equation. In 1977 Kolymbas [30] introduced the first hypoplastic equation for sand. In view of the complexity of the initial model, further research resulted in simpler models with four material constants [31, 32]:

$$\overset{\circ}{\mathbf{T}} = c_1 \frac{1}{2}(\mathbf{TD} + \mathbf{DT}) + c_2\,\mathbf{1}\,\mathrm{tr}\,(\mathbf{TD}) + c_3 \mathbf{T}\sqrt{\mathrm{tr}\,\mathbf{D}^2} + c_4 \frac{\mathbf{T}^2}{\mathrm{tr}\,\mathbf{T}}\sqrt{\mathrm{tr}\,\mathbf{D}^2} \qquad (1.1)$$

Wu & Kolymbas [67] improved equation 1.1 with:

$$\overset{\circ}{\mathbf{T}} = c_1(\mathrm{tr}\,\mathbf{T})\mathbf{D} + c_2 \frac{\mathrm{tr}\,(\mathbf{TD})}{\mathrm{tr}\,\mathbf{T}}\mathbf{T} + c_3 \frac{\mathbf{T}^2}{\mathrm{tr}\,\mathbf{T}}\sqrt{\mathrm{tr}\,\mathbf{D}^2} + c_4 \frac{\mathbf{T}^{*2}}{\mathrm{tr}\,\mathbf{T}}\sqrt{\mathrm{tr}\,\mathbf{D}^2} \qquad (1.2)$$

$\mathbf{T}^* = \mathbf{T} - \frac{1}{3}\mathrm{tr}\,\mathbf{T}\,\mathbf{1}$ is the the deviatoric stress.

---

[1]Often density is taken into account by means of the void ratio e.

Wu & Bauer [65] considered also density (or void ratio) by means of $I_e := (1 - a)\frac{e - e_d}{e_c - e_d} + a$:

$$\overset{\circ}{\mathbf{T}} = c_1(\mathrm{tr}\,\mathbf{T})\mathbf{D} + c_2 \frac{\mathrm{tr}\,(\mathbf{TD})}{\mathrm{tr}\,\mathbf{T}}\mathbf{T} + \left[c_3 \frac{\mathbf{T}^2}{\mathrm{tr}\,\mathbf{T}}\sqrt{\mathrm{tr}\,\mathbf{D}^2} + c_4 \frac{\mathbf{T}^{*2}}{\mathrm{tr}\,\mathbf{T}}\sqrt{\mathrm{tr}\,\mathbf{D}^2} \right] I_e \quad (1.3)$$

The minimum void ratio e_d and a are constants, e_c is the stress-dependent critical void ratio.

Hypoplastic models consist of a linear[2] $\mathcal{L}$ and nonlinear[2] $\mathbf{N}$ term with respect to stretching $\mathbf{D}$. Therefore the following notation is often used:

$$\overset{\circ}{\mathbf{T}} = \mathcal{L} : \mathbf{D} + I_e \mathbf{N}|\mathbf{D}| \qquad (1.4)$$

Wu & Bauer [66] modified equation 1.4 and extended it with a so-called stiffness factor $I_s = (e_{l0}/e)^\beta$:

$$\overset{\circ}{\mathbf{T}} = I_s \left[\mathcal{L} : \mathbf{D} + I_e \mathbf{N}|\mathbf{D}| \right] \qquad (1.5)$$

with $I_e := \left(\frac{e - e_d}{e_c - e_d} \right)^\alpha$. α, β and e_{l0} are material constants.

Further development of hypoplasticity was presented by Gudehus [16] and Bauer [4], often written in the following form:

$$\overset{\circ}{\mathbf{T}} = f_s \mathcal{L} : \mathbf{D} + f_s f_d \mathbf{N}|\mathbf{D}| \qquad (1.6)$$

f_s and f_d are the barotropy[3] and pyknotropy[4] factors, further developed from I_e and I_s. $\mathcal{L}$ and $\mathbf{N}$ from equation 1.6 were introduced by Bauer [3].

A similar development of the type shown in equation 1.6 was presented by von Wolffersdorff [64]. The yield criterion by Matsuoka and Nakai is included in the von Wolffersdorff model. Niemunis & Herle [54] took into account deformation history by developing the so-called *intergranular strain concept*. Adaptions to capture clay behaviour were made by Herle & Kolymbas [24] and Mašín [44]. As the von Wolffersdorff model [64], the model by Mašín [44] includes the Matsuoka-Nakai failure condition.

Barodesy, introduced by Kolymbas [36], is a new constitutive frame for soil and other granular materials. As hypoplastic models, barodesy is conceived as an evolution equation of the type $\overset{\circ}{\mathbf{T}} = \mathbf{h}(\mathbf{T}, \mathbf{D}, e)$. The basic idea is the asymptotic behaviour of granular materials. Common concepts of Soil Mechanics, such as critical states, barotropy[3], pyknotropy[4] and a stress-dilatancy relation are comprised.

---

[2]$\mathcal{L}$ denotes a fourth-order tensor, $\mathbf{N}$ is a second-order tensor.

[3]Barotropy denotes the dependence of stiffness and strength on the stress level.

[4]Pyknotropy denotes the dependence of stiffness and strength on density.

The publications of Fellin & Ostermann [10], Kolymbas [37, 38, 39], Medicus et al. [49] concern further developments and the mathematical structure of barodesy. Barodesy predicts loading paths at medium to large strains, but when it comes to small strain stiffness, barodesy has the same shortcomings as other hypoplastic models of the type $\overset{\circ}{\mathbf{T}} = \mathbf{h}(\mathbf{T}, \mathbf{D}, e)$. Monotonic loading and unloading can be described by the presented model. Considering only the actual void ratio e, the actual effective stress $\mathbf{T}$ and the direction of loading $\mathbf{D}$ it is not possible to fully take into account the deformation history. Furthermore, the description of the hysteretic behaviour at cyclic loading paths is problematic.[5] The model is limited also to stress levels below grain breakage, as a change in particle size distribution is not considered in the model. Rate effects are not taken into account.

The full set of equations of barodesy (for clay) is presented in Table 5.1 on page 70 with the material constants presented in Tables 5.2 and 5.3 (page 73).

1.2 Motivation - Why another constitutive equation?

From the selection of hypoplastic models presented in Section 1.1 it can be seen, that there are already numerous hypoplastic constitutive models. Many models provide realistic results of basic soil features. So why do we need another constitutive model?

Constitutive modelling is a core subject in the field of geotechnical engineering, as the quality of every numerical simulation depends on the used constitutive model. A new approach is an opportunity to create a new mathematical structure. In order to judge the power of a model, it needs to be comprehensible. Interpretable results are generally obtained with a clear mathematical formulation.

Barodesy is characterized by mathematical simplicity. A simple mathematical structure offers potential to further improvement and to assess limitations of the model. The conceptual ideas of barodesy are based on well accepted soil features and the resulting mathematical model is simple. Barodesy relates to physical concepts. It was therefore possible to include previous geotechnical theories in the structure of the model. Another requirement was to develop a calibration procedure, which is based on standard laborartory tests. It was important that the material constants of barodesy are related to well-known parameters of Soil Mechanics.

Research in the field of constitutive modelling does not only create new models, but also trains the ability to understand soil behaviour and to judge in which cases assumptions and simplifications are acceptable.

---

[5]These restrictions apply to all models of the type $\overset{\circ}{\mathbf{T}} = \mathbf{h}(\mathbf{T}, \mathbf{D}, e)$.

About the name barodesy   Kolymbas explains why he introduced a new name (*barodesy*) as follows [38]: *"One should be cautious with introducing new names. Too many neologisms create confusion. However, sometimes new names are needed to denote new notions. There is an abundance of elastic and plastic concepts equipped with prefixes such as hypo-, para-, hyper-, and others. Therefore, the author suggests to avoid using the words elasticity and plasticity (to the extent the latter is associated with notions such as yield surface, elastic regime etc., originally created for metals), because they are not the only framework to describe granular materials such as soil. ..."*

1.3   Symbols and Notations

In this thesis the symbolic notation is used for stress $\mathbf{T}$ and stretching $\mathbf{D}$, but in some cases the more familiar symbol σ_i instead of T_i is used for the principal stresses. Note that for σ_i compression is defined positive, for the principal components of $\mathbf{T}$ compression is defined negative. Tensors are written in bold capital letters (e.g. $\mathbf{X}$). $|\mathbf{X}| := \sqrt{\operatorname{tr} \mathbf{X}^2}$ is the Euclidean norm of $\mathbf{X}$, $\operatorname{tr} \mathbf{X}$ is the sum of the diagonal components of $\mathbf{X}$. The superscript 0 marks a normalised tensor, i.e. $\mathbf{X}^0 = \mathbf{X}/|\mathbf{X}|$. Stresses are considered as effective ones, the normally used dash is omitted. As in this work only rectilinear extensions are examined, the co-rotational, objective stress rate[6] $\overset{\circ}{\mathbf{T}}$ is replaced with $\dot{\mathbf{T}}$. The stretching tensor $\mathbf{D}$ is the symmetric part of the velocity gradient.[7] The following abbreviations are used: $\delta := \operatorname{tr} \mathbf{D}^0, \dot{\varepsilon} := |\mathbf{D}|, \sigma := |\mathbf{T}|$. The void ratio e is the ratio of the volume of the voids V_p to the volume of the solids V_s. $p := -\frac{1}{3}\operatorname{tr} \mathbf{T}$ is the mean effective stress, $\varepsilon_{\mathrm{vol}} = \operatorname{tr} \varepsilon$ is the volumetric strain.[8]

For axisymmetric conditions often the Rendulic plane is used, see Figure 1.1. For a conventional triaxial compression or oedometric compression test the axial stress is denoted with σ_1 and the radial stress is denoted with $\sigma_2(= \sigma_3)$. The associated strains are ε_1 and $\varepsilon_2 = \varepsilon_3$. Symbols used in Critical State Soil Mechanics are $q := \sigma_1 - \sigma_3$ and $\varepsilon_q := 2/3 \cdot (\varepsilon_1 - \varepsilon_3)$. Note that several definitions for dilatancy can be found. In this work $\delta = \operatorname{tr} \mathbf{D}^0 = -\dot{\varepsilon}_{\mathrm{vol}}/\dot{\varepsilon}$, $\tan \beta = \dot{\varepsilon}_{\mathrm{vol}}/-\dot{\varepsilon}_1$ and the so-called *strain increment ratio* $m = \dot{\varepsilon}_{\mathrm{vol}}/\dot{\varepsilon}_q$ are used for the formulations of different *stress-dilatancy* relations (see Section 2.1.4, page 20). δ can be expressed by means of $\tan \beta$ as follows: From $\tan \beta = \dot{\varepsilon}_{\mathrm{vol}}/ - \dot{\varepsilon}_1 = (-\dot{\varepsilon}_1 - 2\dot{\varepsilon}_2)/\dot{\varepsilon}_1$ we get $\dot{\varepsilon}_2 =$

---

[6]The term objectivity points to the fact that material behaviour is frame indifferent, i.e. the behaviour is independent of the observers motion.

[7]In general, stretching $\mathbf{D}$ is only approximately equivalent to the strain rate $\dot{\varepsilon}$. For rectilinear extensions, $\mathbf{D}$ equals $\dot{\varepsilon}$.

[8]For compressive strain, ε_i is defined positiv.

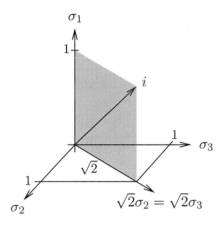

Figure 1.1: Rendulic plane; i refers to isotropic compression, i.e. $\sigma_1 = \sigma_2 = \sigma_3$

$-\dot{\varepsilon}_1(1 + \tan\beta)/2$. With $\dot{\varepsilon}_2 = -\dot{\varepsilon}_1(1 + \tan\beta)/2$ in $\delta = (\dot{\varepsilon}_1 + 2\dot{\varepsilon}_2)/\sqrt{\dot{\varepsilon}_1^2 + 2\dot{\varepsilon}_2^2}$ yields:

$$\delta = \frac{\tan\beta}{\sqrt{1 + (1 + \tan\beta)^2/2}} \tag{1.7}$$

m can be expressed by means of $\tan\beta$ as follows:

$$m = \frac{\dot{\varepsilon}_{\text{vol}}}{\dot{\varepsilon}_q} = -\frac{3\tan\beta}{3 + \tan\beta} \tag{1.8}$$

Note that $m = \dot{\varepsilon}_{\text{vol}}/\dot{\varepsilon}_q$ applies only for axisymmetric conditions. Using invariants of $\mathbf{D}^0$ we get $m = -\operatorname{tr}\mathbf{D}^0/(2/3\sqrt{3J_2})$ with

$$J_2 := (\operatorname{tr}\mathbf{D}^0)^2/3 - 1/2 \left[\underbrace{(\operatorname{tr}\mathbf{D}^0)^2 - \operatorname{tr}\left(\mathbf{D}^{02}\right)}_{1} \right] = \frac{3 - \operatorname{tr}\mathbf{D}^{02}}{6} \tag{1.9}$$

It follows that m can be expressed as follows:

$$m = \frac{-3\delta}{\sqrt{6 - 2\delta^2}} \tag{1.10}$$

Chapter 2

Soil behaviour

In this chapter clay and sand characteristics, based on the present state of research are summarised. The aspects presented here will be used for modelling and calibrating barodesy in the subsequent chapters. In particular, common properties and differences between sand and clay are addressed. Apart from sand and clay, nature provides different types of soil, such as silt and mixtures thereof. Sand and clay in their pure form are often investigated in the laboratories. It is important to distinguish between *natural soil* and artificially prepared *reconstituted soil*. Natural soil is exposed to aging. Physical (e.g. deposition and erosion) and chemical (e.g. cementatation) effects influence the soil structure. Various authors (e.g. Leroueil et al. [41], Callisto & Calabresi [7]) discuss the differences of natural and reconstituted soil. The soil properties presented in this Chapter refer to soils, reconstituted in the laboratory.

2.1   Soil properties, qualitative and quantitative description

The main differences between sand and clay lie in the size of grains and the specific surface, cf. Figure 2.1. Specific surfaces of clay are in the order of several square meters per gram, whereas specific surfaces of sand lie in the range of several 10^{-4} m^2/g. Concerning the (hydrostatic) compression behaviour, the compressibility of clay is much higher than of sand, cf. Figure 2.2. The data for loose (*host-l-iso*) and dense (*host-d-iso*) sand refer to Hostun Sand[1] from Desrues et al. [9], the clay data refer to London Clay and are from Gasparre [13].

2.1.1   Critical State Concept

Regarding strength, we find that so-called *critical strength* of clay is in general lower than the critical strength of sand, cf. Figure 2.3. Again, the data for sand refer to CU-test of Hostun Sand [9], the clay data refer to London Clay [13]. The clay

---

[1]Hostun is a commune in the Drôme department in the French alps. Hostun RF sand is a siliceous sand with the following parameters: The coefficient of uniformity $D_{60}/D_{10} = 1.7$, the minimum void ratio lies between 0.624 and 0.648, the maximum void ratio between 0.961 and 1.041 [9].

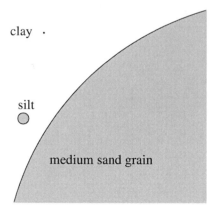

Figure 2.1: Scale-up of typical grains of sand, silt and clay

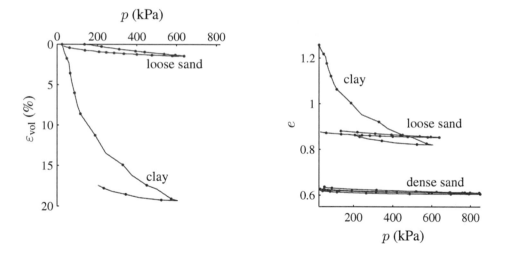

Figure 2.2: Normally consolidated clay is softer than loose sand. The experimental results shown here refer to hydrostatic compression tests, data from Desrues et al. [9] (for sand) and from Gasparre [13] (for clay).

data in Figure 2.3(b) refer to Dresden clay.[2] At critical states, the stiffness vanishes ($\dot{q} = 0, \dot{p} = 0$) and ongoing shear deformation occurs under isochoric deformation. Under these conditions stress level, shearing resistance and void ratio are linked. The critical/residual shear strength depends on the stress level (i.e. the mean stress p) and is given through the so-called critical state line (CSL), cf. Figure 2.4(a). The

---

[2]The data of Dresden Clay are taken from the master's thesis of Katharina Bergholz: *Experimentelle Bestimmung von nichtlinearen Spannungsgrenzbedingungen*, Technische Universität Dresden, 2009.

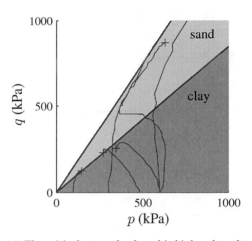

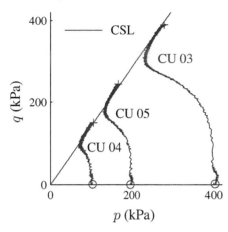

(a) The critical strength of sand is higher than the critical strength of clay. The experimental results refer to CU tests, by Desrues et al. [9] and Gasparre [13].

(b) The stress paths of CU-tests approach the CSL. Experimental results obtained with Dresden clay.

Figure 2.3: Critical strength of soil

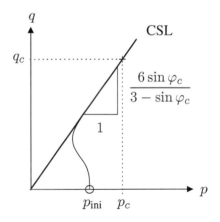

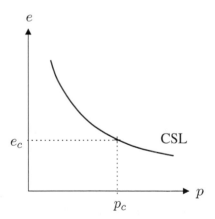

(a) The critical state line is related to the critical friction angle φ_c.

(b) The higher the mean stress p is, the lower is the void ratio at critical states.

Figure 2.4: Critical State line (CSL)

higher the mean stress p is, the lower is the void ratio at critical states e_c, cf. Figure 2.4(b). The CSL in Figure 2.4 is represented in q vs. p and in e vs. p diagrams.

The Critical State Concept is valid for sand and clay. Many constitutive models for sand and clay refer to the Critical State Concept, e.g. the Modified Cam Clay by Roscoe & Burland [58], the hypoplastic model for clay by Mašín [44], the Severn-Trent-Sand model by Gajo & Muir Wood [11], the hypoplastic model for sand by von Wolffersdorff [64]. The critical state line (CSL) in the p - q plot is given through the critical friction angle φ_c (Figure 2.4a).

The critical strength is characterised through the critical friction angle φ_c, which can be determined, for example, by an undrained triaxial compression test, cf. experimental results in Figure 2.3(b) and Figure 2.4(a). The slope of the critical state line for compression in the p - q plane is $q/p = M = 6 \sin \varphi_c/(3 - \sin \varphi_c)$. From a critical stress state, φ_c can be obtained.

$$\sin \varphi_c = \left(\frac{3q}{q + 6p} \right)_c \tag{2.1}$$

2.1.2 Stiffness

Oedometric stiffness: Soil stiffness depends on the stress-level and on density. The higher stress and density are, the stiffer is the soil. For oedometric compression, incremental stiffness can be denoted as $E_s = \dot{\sigma}_1/\dot{\varepsilon}_1$. Muir Wood [51] expresses the relation introduced by Ohde [55] and Janbu [27] as follows:

$$E_s \propto \sigma^* \left(\frac{\sigma}{\sigma^*} \right)^\alpha \quad \text{or} \quad \frac{E_s}{\sigma^*} = \chi \left(\frac{\sigma}{\sigma^*} \right)^\alpha \tag{2.2}$$

α and χ are material constants, σ^* is a reference stress. Figure 2.5(a) (slightly modified from Muir Wood [51]) shows values of the exponent α for various soils. If $\alpha = 0$, the stiffness is constant.[3] Note that for $\alpha = 1$ a linear relation in the $\ln \sigma_1$ - ε_1 plot follows. $\alpha = 1$ provides realistic results for clays. Figure 2.5(b) shows ranges for χ-values according to equation 2.2 for different soils. $\chi \cdot \sigma^*$ is the magnitude of the stiffness at the stress level σ^*.

Figure 2.6 shows oedometric compression curves of Hostun Sand. The loading behaviour is approximated with equation 2.2. As a reference stress σ^*, 100 kPa is chosen according to Muir Wood [51], in order to compare the parameter χ in Figure 2.5(b) with the parameters chosen in Figure 2.6. Note that any other value σ^* can be assumed. σ^* has no influence on the exponent α, but on the magnitude of the modulus number χ. In Figure 2.6(a) χ and α are chosen in order to obtain the best fit. This yields different exponents for the loose (*host-l-oed*) and dense (*host-d-oed*) sample, cf. Figure 2.5(a). Jänke [28] sums up the influences on α as follows: The

---

[3]It follows a linear relation in the σ_1 - ε_1 plot. For rock $\alpha = 0$ is realistic according to Muir Wood.
[4]Following Muir Wood [51] the notation for preloaded soil is also used for sand.

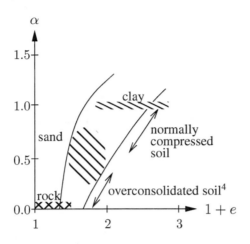

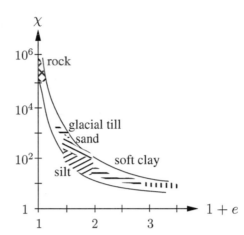

(a) Stiffness exponent α for various soils

(b) So-called modulus number χ for different soils: the higher void ratio is, the softer is soil

Figure 2.5: Visualisation of the parameters of equation 2.2. Figures are slightly modified from Muir Wood [51] based on Janbu [27].

rougher the grains are, the lower is the exponent α. The higher the relative density is, the lower is the exponent α, see Figure 2.6(a). Setting α equal to 0.5, gives an approximation of the compression behaviour of Hostun Sand [9], cf. Figure 2.6(b). The approximation gives reasonable results, only for certain stress-, strain- and void ratio ranges.

Shear stiffness: One can express the stiffness of an undrained triaxial compression test (cf. Figure 2.7) by the shear modulus G:

$$G = \frac{1}{3}\frac{\partial q}{\partial \varepsilon_q} \tag{2.3}$$

Atkinson [1] describes how the shear modulus G at small strains is related to mean stress p and overconsolidation ratio[5] $OCR = p_0/p_{\text{ini}}$:

$$\frac{G}{p} = A \cdot OCR^n \tag{2.4}$$

Equation 2.4 is illustrated in Figure 2.8. G/p reduces with ongoing shear strain. For normally consolidated soil $G/p = A$, as the overconsolidation ratio equals 1, cf. the lowest curve in Figure 2.8.

---

[5]How p_0 is defined is shown in Figure 2.24(b) on page 29. Soil is isotropically normally compressed up to p_0 and then unloaded to p_{ini}.

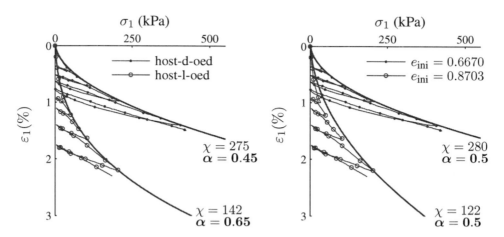

(a) Compression curves are fitted with parameters χ and α.

(b) Exponent α is set to 0.5.

Figure 2.6: Oedometric compression curves can be approximated by equation 2.2. The data (host-d-oed, host-l-oed) refer to oedometric tests of dense and loose Hostun sand [9].

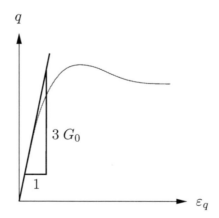

Figure 2.7: The initial shear stiffness G_0 is defined in the q - ε_q plot and can be e.g. obtained by a triaxial test.

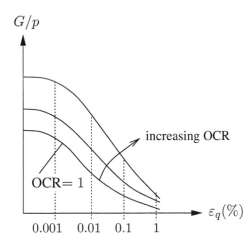

Figure 2.8: The normalized shear stiffness G/p depends on the OCR. Figure modified from Atkinson [1];

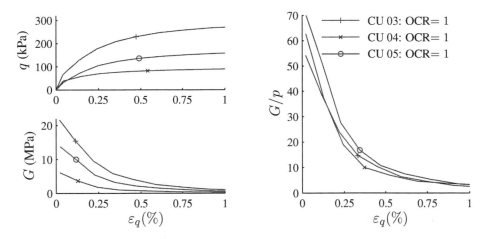

(a) Shear stiffness G reduces with ongoing shear strain.

(b) G is normalized with p. According to relation 2.4, test results with the same OCR should coincide.

Figure 2.9: Shear stiffness under undrained shear of Dresden Clay samples

In Figure 2.9 CU-test results with Dresden Clay are shown. In Figure 2.9(b) the shear stiffness G is normalized with p. According to relation 2.4, test results with the same OCR should coincide.

Isotropic compression	Oedometric compression	Critical state
$\varepsilon_1 : \varepsilon_2 : \varepsilon_3 = 1$	$\varepsilon_1 = \text{const}$ $\varepsilon_2 = \varepsilon_3 = 0$	$\varepsilon_1 : \varepsilon_2 : \varepsilon_3 = \text{const}$ $\varepsilon_1 + \varepsilon_2 + \varepsilon_3 = 0$
$\sigma_1 : \sigma_2 : \sigma_3 = 1$	$\sigma_2 : \sigma_1 = K_0$ $\sigma_2 = \sigma_3$	$\sigma_2 : \sigma_1 = K_c$ $\sigma_2 = \sigma_3$

Table 2.1: Ratios of principal strains and principal stresses for certain axisymmetric loadings

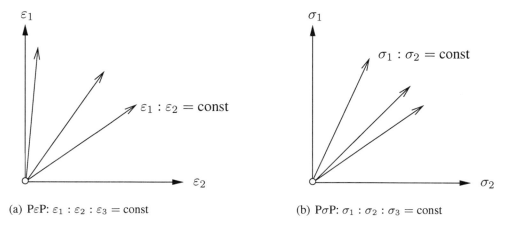

(a) PεP: $\varepsilon_1 : \varepsilon_2 : \varepsilon_3 = \text{const}$ (b) PσP: $\sigma_1 : \sigma_2 : \sigma_3 = \text{const}$

Figure 2.10: Proportional strain paths are characterized by $\varepsilon_1 : \varepsilon_2 : \varepsilon_3 = \text{const}$. In the same sense, proportional stress paths are characterized by $\sigma_1 : \sigma_2 : \sigma_3 = \text{const}$.

2.1.3 Proportional paths and Goldscheider's rules

Proportional strain paths PεPs are paths with constant ratios of the principal strains (e.g. the strain paths of isotropic compression, oedometric compression and isochoric compression), i.e. $\varepsilon_1 : \varepsilon_2 : \varepsilon_3 = \text{const}$, cf. Figure 2.10(a). In the same sense, paths with constant ratios of the principal stresses are called *proportional stress paths* PσPs, i.e. $\sigma_1 : \sigma_2 : \sigma_3 = \text{const}$, cf. Figure 2.10(b). Some typical PεPs and PσPs are summarized in Table 2.1. Note that also the term *asymptotic states*[6], introduced by Gudehus et al. [19], is often used for *proportional paths* (PPs). The *asymptotic state concept* refers to the Goldscheider rules described below.

Goldscheider [14] applied proportional strain paths to sand samples and observed the corresponding stress paths. He formulated two rules, based on the results of true triaxial tests, see Figure 2.11.

---

[6]Deforming a sample with constant stretching, an asymptotic state is reached when the stress ratio stays constant.

[7]Institute of Soil Mechanics and Rock Mechanics, University of Karlsruhe.

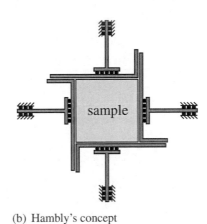

(a)   (b) Hambly's concept

Figure 2.11: True Triaxial Test[7]

Goldscheider's first rule: Starting at the stress free state, $\mathbf{T} = \mathbf{0}$, proportional strain paths lead to proportional stress paths, see Figures 2.12(a) and 2.12(b). The directions of proportional stress paths PσP are related to specific stretchings $\mathbf{D}$ (i.e. the directions of proportional strain paths PεP). Gudehus & Mašín [18] propose a graphical representation of how PσPs and PεPs are linked. They consider the angles $\psi_{\dot\varepsilon}$ and ψ_σ according to Figure 2.13, in order to specify the directions of proportional paths in the Rendulic plane and to create a graphical representation of the relation between proportional strain and stress paths. A sketch of a $\psi_{\dot\varepsilon}$ - ψ_σ plot is shown in Figure 2.14(a). The directions of PσPs are here assumed to be independent of the initial void ratio. $\psi_{\dot\varepsilon}$ and ψ_σ are linked by the following requirements, cf. Table 2.2:

- All proportional stress paths lie in the compression area, i.e. $-35.3° < \psi_\sigma < 54.7°$.

- For critical states, i.e. $\psi_{\dot\varepsilon} = \pm 90°$, ψ_σ depends on the critical friction angle φ_c, cf. Table 2.2.

- For hydrostatic compression/extension, i.e. $\psi_{\dot\varepsilon} = 0°/180°$, it follows that ψ_σ equals $0°$. Note that isotropic extension is not considered by Gudehus & Mašín [18], but is here mentioned for the sake of completeness and is used in Chapter 4.

The relation between $\psi_{\dot\varepsilon}$ and ψ_σ for the directions d and $-d$ according to Table 2.2 is a suggestion by Gudehus & Mašín [18], cf. also Figures 2.13 and 2.14(a).

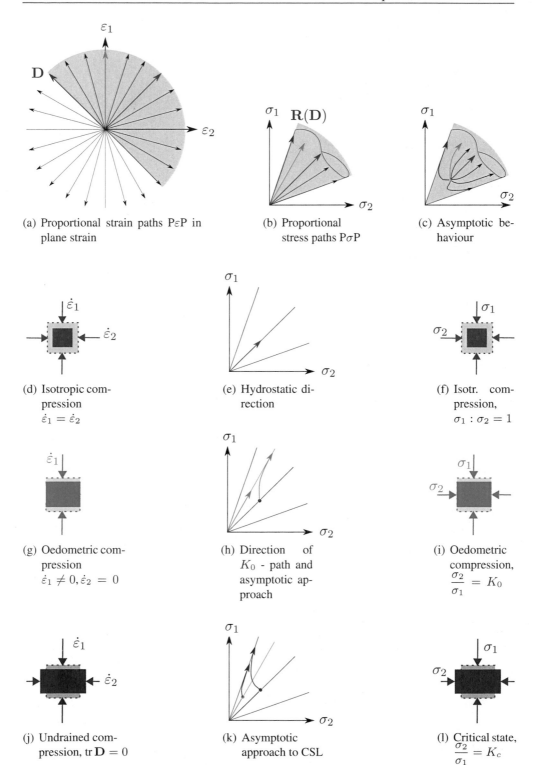

(a) Proportional strain paths PεP in plane strain

(b) Proportional stress paths PσP

(c) Asymptotic behaviour

(d) Isotropic compression
$\dot{\varepsilon}_1 = \dot{\varepsilon}_2$

(e) Hydrostatic direction

(f) Isotr. compression,
$\sigma_1 : \sigma_2 = 1$

(g) Oedometric compression
$\dot{\varepsilon}_1 \neq 0, \dot{\varepsilon}_2 = 0$

(h) Direction of K_0 - path and asymptotic approach

(i) Oedometric compression,
$\dfrac{\sigma_2}{\sigma_1} = K_0$

(j) Undrained compression, tr $\mathbf{D} = 0$

(k) Asymptotic approach to CSL

(l) Critical state,
$\dfrac{\sigma_2}{\sigma_1} = K_c$

Figure 2.12: Proportional paths and asymptotic behaviour (schematically)

Processes and states		$\psi_{\dot{\varepsilon}}$	ψ_σ
Isotropic compression	i	$0°$	$0°$
Critical state	c	$90°$	$\arctan\left(1/(\sqrt{2}K_c)\right) - \arctan\left(1/\sqrt{2}\right)$
Critical state	$-c$	$-90°$	$\arctan\left(K_c/\sqrt{2}\right) - \arctan\left(1/\sqrt{2}\right)$
	d	$144.7°$	$54.7°$
	$-d$	$-125.3°$	$-35.3°$
Isotropic extension	$-i$	$180°$	$0°$

Table 2.2: Examples for $\psi_{\dot{\varepsilon}}$ and ψ_σ for isotropic compression i, critical states (c, $-c$) and for d and $-d$, cf. Figures 2.13 and 2.14(a).[9] The relation between $\psi_{\dot{\varepsilon}}$ and ψ_σ, for the directions d and $-d$ is merely a suggestion. Isotropic extension ($-i$) is not suggested by Gudehus & Mašín [18], but is here mentioned for the sake of completeness and is used in Chapter 4.

Goldscheider's second rule: Starting at $\mathbf{T} \neq \mathbf{0}$, proportional strain paths lead asymptotically to the corresponding proportional stress paths starting at $\mathbf{T} = \mathbf{0}$, see Figure 2.12(a) and 2.12(c). This means that proportional stress paths act as attractors. Goldscheider's rules are illustrated by the following examples: Hydrostatic compression, i.e. $\mathbf{D} = \text{const} = -\mathbf{1}$, causes a hydrostatic stress path with $\sigma_1 : \sigma_2 : \sigma_3 = 1$, cf. Figures 2.12(d)-(f). In the same sense, oedometric deformation ($\mathbf{D} = \text{const}$) leads to a proportional stress path with $\sigma_1 : \sigma_2 : \sigma_3 = \text{const}$ (i.e. $\sigma_2 : \sigma_1 = K_0$), cf. Figures 2.12(g)-(i) and Table 2.1. Applying isochoric deformation ($\text{tr}\,\mathbf{D} = 0$) starting from oedometric, hydrostatic or any other admissible state will cause asymptotic approaches to the critical state line, cf. Figure 2.12(k). The consolidated undrained (CU) triaxial test is an example of the two Goldscheider rules: Experimental results which confirm the two Goldscheider rules were carried out for sand and clay by Chu & Lo [8], Goldscheider [14], Topolnicki [62], Topolnicki et al. [63], cf. Figure 2.15. Figure 2.15(a) shows stress paths obtained by true triaxial tests starting from different consolidation pressures. Deforming the samples with constant stretching $-\dot{\varepsilon}_{\text{vol}}/\dot{\varepsilon}_1 = 0.11$, the same proportional stress ratio is reached regardless of the initial consolidation pressure. Figure 2.15(b) shows different loading, unloading and reloading cycles of a true triaxial test. Under long enough proportional loading by a PεP (respectively reloading), the stress paths reach the proportional stress path which corresponds to $-\dot{\varepsilon}_{\text{vol}}/\dot{\varepsilon}_1 = 0.11$.

Starting from states with $\mathbf{T} \neq \mathbf{0}$ and applying only a unit strain increment[10] $\Delta\varepsilon$ in different directions of loading (cf. Figure 2.13a), leads to different directions and values of stress increments, which depend on the initial stress state, deformation history and initial density.[11] The envelope of all stress increments is called *response envelope* (Gudehus [15]), cf. Figure 2.14(b).

---

[9]If only compressive stress states are permitted, it follows that ψ_σ is delimited as follows: $-\arctan(1/\sqrt{2}) \leq \psi_\sigma \leq -\arctan(1/\sqrt{2}) + 90°$ (i.e. $-35.26° \leq \psi_\sigma \leq 54.74°$).

[10]$\Delta\varepsilon = \sqrt{\dot{\varepsilon}_1^2 + 2\dot{\varepsilon}_2^2}\,\Delta t = 1$

[11]The lower the void ratio is, the larger is the stress increment to obtain a unit strain increment.

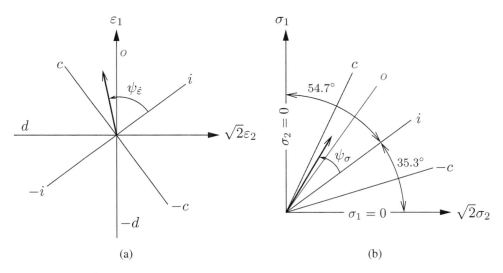

Figure 2.13: Definition of $\psi_{\dot{\varepsilon}}$ and ψ_σ according to Gudehus & Mašín [18]: i refers to isotropic compression, $-i$ to isotropic extension, c to isochoric triaxial compression and $-c$ to isochoric triaxial extension.

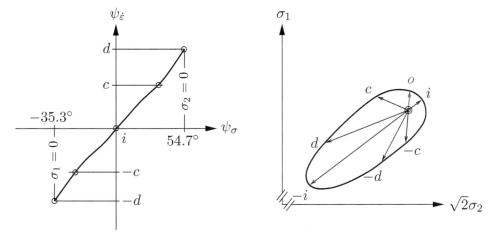

(a) A qualitative sketch of a reasonable relation between $\psi_{\dot{\varepsilon}}$ and ψ_σ according to Gudehus & Mašín [18]. The requirements from Table 2.2 are met. *Caution:* Not a constant scaling is used in this Figure.

(b) A *response envelope* comprises all stress increments for different loading directions obtained by a constant strain increment.

Figure 2.14: Directions of stress paths obtained by different directions of stretching (schematically)

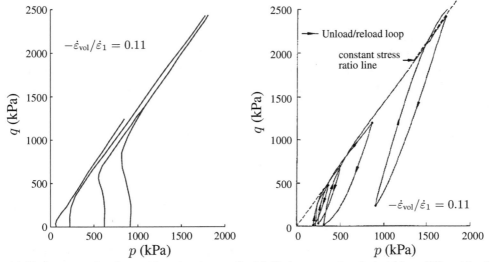

(a) Under proportional stretching, the stress paths approach approximately the same stress ratio regardless of the initial consolidation pressure.

(b) Under proportional stretching, different loading paths lead to the same stress ratio line.

Figure 2.15: True triaxial test results with Sydney sand obtained with constant stretching $-\dot{\varepsilon}_{\mathrm{vol}}/\dot{\varepsilon}_1 = 0.11$. Chu & Lo [8] illustrate the asymptotic approach to a certain stress ratio. The loading histories of the samples differ. In (a) the stress paths start from different confining pressures; in (b) the sample is loaded with several loading cycles.

2.1.4   Stress-dilatancy relations

The mobilised friction angle φ_m is related to dilatancy $\tan\beta := -\dot\varepsilon_{\text{vol}}/\dot\varepsilon_1$.[12] If soil is dilatant (i.e. $\tan\beta > 0$), the mobilised friction angle is larger than the critical friction angle. According to Taylor [61], strength is linked to dilatancy.

Rowe's relation: According to Rowe [59] stress is linked to dilatancy as follows, cf. Muir Wood [50]:[13]

$$\frac{\sigma_1}{\sigma_2} = -\frac{2\dot\varepsilon_2}{\dot\varepsilon_1} \cdot \frac{1 + \sin\varphi_c}{1 - \sin\varphi_c} \tag{2.5}$$

Chu & Lo's relation: Chu & Lo [8] found an empirical relationship between stress ratio and dilatancy $\tan\beta$. They observed that the stress ratios q/p of proportional stress paths obtained by proportional strain paths are related to the corresponding dilatancies $\tan\beta$ as shown in Figure 2.16(a) and (b).[14]

Connecting isotropic compression and the critical state according to Table 2.3 with a straight line in the q/p vs. $\tan\beta$ plot, leads to the following linear relation between q/p and $\tan\beta$ (Figure 2.16 b):

$$\frac{q}{p} = \frac{M}{3}\tan\beta + M \qquad \text{with} \quad M = \frac{6\sin\varphi_c}{3 - \sin\varphi_c} \tag{2.6}$$

Figure 2.15 shows triaxial test results obtained with a constant dilatancy strain path, i.e. $\tan\beta = 0.11$. In Figure 2.17(b) experimental data from Chu & Lo [8] are compared with equation 2.6. Figure 2.17(a) shows q/p-ε_1 plots of triaxial tests obtained with constant stretchings. The test results are marked in Figure 2.17(b). Conventional drained (CD) triaxial test results of Hostun Sand[15] from Desrues et al. [9] are shown in Figure 2.18(a). The higher the dilatancy at peak is, the higher is the peak strength. Peaks of CD-tests can also be described by equation 2.6, cf. Figure 2.18(b).[16]

---

[12]E.g. for oedometric compression is $\varphi_m = \arcsin\dfrac{\sigma_1 - K_0\sigma_1}{\sigma_1 + K_0\sigma_1}$ and $\tan\beta = -1$.

[13]Equation 2.5 applies for triaxial compression. σ_1 is the axial stress, σ_2 the radial stress, $\dot\varepsilon_1$ and $-\dot\varepsilon_2$ are the respective strain rates. In Figure 4.4(b) (page 55) Rowe's relation is plotted and compared with other stress-dilatancy relations.

[14]The relation in equation 2.6 captures only states within a certain range of $\tan\beta$. Chu & Lo [8] show experimental accordance in the range $-1 < \tan\beta \lesssim 0.8$, cf. Figure 2.17(b).

[15]The experimental data refer to *host-d-triaxc-cd-100: cdhfd11, host-d-triaxc-cd-300: hfdw08, host-d-triaxc-cd-300: hfdw09*, marked with A, B, C. Figure 2.18(b) is completed with more dense triaxial test data from Desrues et al. [9]. The critical friction angle φ_c is 33.8°.

[16]Note that Chu & Lo [8] developed their relation for proportional paths. It is assumed in this thesis and by e.g. Mašín [46] that peak states of CD tests and proportional paths coincide.

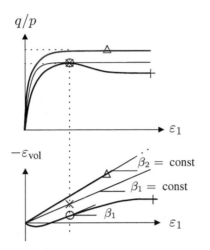

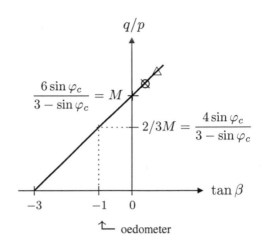

(a) Schematic diagram of triaxial tests with proportional stretchings ($\triangle$, $\times$) and a conventional CD test ($\bigcirc$).

(b) Stress ratios of proportional stress paths q/p vs. dilatancy $\tan \beta$, according to Chu & Lo [8]: The higher $\tan \beta$ is, the higher is the obtained stress ratio ($\triangle$, $\times$).

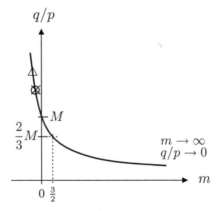

(c) Stress-ratio q/p, strain increment ratio m, according to Luo et al. [42]: The lower m is, the higher is the obtained stress ratio ($\triangle$, $\times$) and q/p at the peak ($\bigcirc$);

Figure 2.16: Figures (b) and (c) show the stress-dilatancy relation, according to Chu & Lo [8] and Luo et al. [42]. Proportional paths can be described by equations 2.6 and 2.8, cf. Table 2.3. The tests according to (a) are marked in (b) and (c). Note that Chu & Lo and Luo et al. developed their relation for proportional paths. In this thesis it is assumed that peak states of CD tests ($\bigcirc$) and proportional paths ($\times$) coincide, cf. Figure 2.18(b) and e.g. Mašín [46].

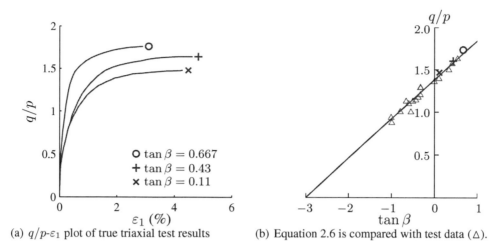

(a) q/p-ε_1 plot of true triaxial test results

(b) Equation 2.6 is compared with test data ($\triangle$).

Figure 2.17: Experimental data with Sydney sand from Chu & Lo [8] obtained by triaxial and true triaxial tests with constant stretchings. The constant M is 1.38 ($\varphi_c \approx 34.1°$). The tests from Figure (a) are marked in (b).

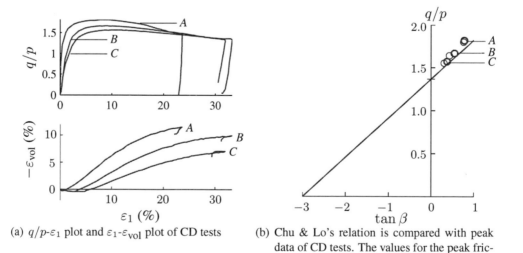

(a) q/p-ε_1 plot and ε_1-ε_{vol} plot of CD tests

(b) Chu & Lo's relation is compared with peak data of CD tests. The values for the peak friction angle and dilatancy at the peak are taken from Desrues et al. [9].

Figure 2.18: Conventional drained triaxial test results of Hostun Sand, data from Desrues et al. [9]. The higher the dilatancy at the peak is, the higher is the peak strength.

Processes and states	δ	$\tan\beta$	m	q/p (equ. 2.6 and 2.8)
Isotropic compression	$-\sqrt{3}$	-3	∞	0
Oedometric compression	-1	-1	$3/2$	$2/3M$
Critical state	0	0	0	M
Peak states	>0	>0	<0	$>M$
Isotropic extension[18]	$\sqrt{3}$	-3	$-\infty$	0

Table 2.3: Dilatancy measures δ, $\tan\beta$, strain increment ratio m and resulting stress ratio q/p with $M = (6\sin\varphi_c)/(3 - \sin\varphi_c)$, cf. Figure 2.16. Jáky's relation $K_0 = 1 - \sin\varphi_c$ for oedometric compression can be rewritten as $q/p = (3\sin\varphi_c)/(3 - 2\sin\varphi_c)$.

Chu and Lo's relation (equation 2.6) also predicts K_0 values, cf. Figures 2.16(b) and 2.19. Equation 2.6 simplifies for oedometric compression ($\tan\beta = -1$) to $\dfrac{q}{p} = \dfrac{2}{3}M = \dfrac{4\sin\varphi_c}{3 - \sin\varphi_c}$, and thus

$$\sigma_2 = K_0\,\sigma_1 \quad \text{with} \quad K_0 = \frac{9 - 7\sin\varphi_c}{9 + 5\sin\varphi_c} \tag{2.7}$$

Jáky's relation[17] $K_0 = 1 - \sin\varphi$ and Chu and Lo's predictions for K_0 are well in accordance, as shown in Figure 2.19, where they are compared with experimental data from Muir Wood [50]. Expressing Jáky's relation as a q/p ratio, yields $(q/p)^{\text{oed}} = 3\sin\varphi_c/(3 - 2\sin\varphi_c)$.

Alternative representation of equation 2.6: In addition to the q/p - $\tan\beta$ - formulation, Luo et al. [42] propose the following formulation, which coincides with equation 2.6:[19]

$$\frac{q}{p} = \frac{3}{3 + m}M \quad \text{with} \quad m = \frac{\dot{\varepsilon}_{\text{vol}}}{\dot{\varepsilon}_q} = -\frac{3\tan\beta}{3 + \tan\beta} \tag{2.8}$$

A more general writing of (the so-called *strain increment ratio*) m gives[20]:

$$m = \frac{-3\delta}{\sqrt{6 - 2\delta^2}} \tag{2.9}$$

---

[17]Muir Wood [50] points out the fact that the K_0 value is dependent on density and is expected to be linked to the peak friction angle. Only for loose sand and normally consolidated clay, φ_c equals φ_p. In this thesis φ_c is used in Jáky's relation, i.e. $K_0 = 1 - \sin\varphi_c$.

[18]Isotropic extension is not representable by equations 2.6 and 2.8, but is here mentioned for the sake of completeness and is used in Chapter 4.

[19]Equation 2.6 characterizes the direction of stretching by $\tan\beta$, equation 2.8 uses m.

[20]For better comprehension of equation 2.9 compare equations 1.8 - 1.10 on page 5.

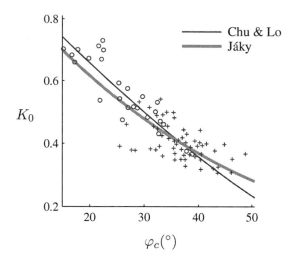

Figure 2.19: Jáky's and Chu & Lo's predictions for K_0 are compared with clay (o) and sand (+) experiments, data from Muir Wood [50].

Figure 2.16(c) shows the m-q/p relation (equation 2.8) and Table 2.3 provides values for m for several processes and states.

Addition for sand: Bolton [5] presents a strength-dilatancy relation for sand. To link $\varphi_p - \varphi_c$ and $\tan \beta_{max}$ with density, he introduced a relative density index I_R:

$$I_R := I_D \left(10 - \ln \frac{p_p}{\sigma^*}\right) - 1 \tag{2.10}$$

with

- the relative density $I_D = \dfrac{e_{max} - e}{e_{max} - e_{min}}$

- the maximum void ratio e_{max}

- the minimum void ratio e_{min}

- the mean effective stress at peak p_p, and the reference stress $\sigma^* = 1$ kPa

The peak friction angle is written as φ_p. Note that e is the void ratio after consolidation and not the void ratio at the peak. Bolton [5] proposes the following empirical relations:

$$\varphi_p - \varphi_c = 3I_R \tag{2.11}$$

$$(\tan \beta)_{max} = 0.3I_R \tag{2.12}$$

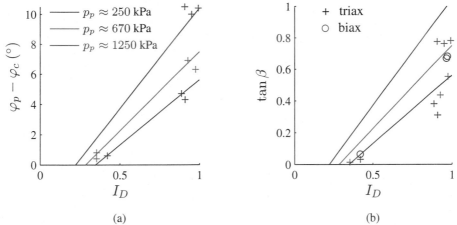

(a)   (b)

Figure 2.20: Bolton's strength-dilatancy relation relates $(\varphi_p - \varphi_c)$ and dilatancy $\tan \beta$ with the relative density I_D. Experimental data from Desrues et al. [9]

Equation 2.11 applies for triaxial tests, equation 2.12 for triaxial and biaxial tests. In Figure 2.20, the relations according to equations 2.11 and 2.12 are plotted for various mean stresses p_p and compared with experimental data from Hostun sand from Desrues et al. [9].

Eliminating I_R from equations 2.11 and 2.12 gives the following relation between φ_p and $(\tan \beta)_{\max}$:

$$\varphi_p = 10(\tan \beta)_{\max} + \varphi_c \tag{2.13}$$

In Figure 2.21 the relation according to equation 2.13 is plotted for different critical friction angles φ_c. The stress rate q/p in Chu & Lo's relation according to equation 2.6 can be expressed as a function of the peak friction angle:

$$\frac{q}{p} = \frac{6 \sin \varphi_p}{3 - \sin \varphi_p} = \frac{M}{3}(\tan \beta)_{\max} + M \tag{2.14}$$

and hence

$$\sin \varphi_p = \frac{3M(\tan \beta)_{\max} + 9M}{M(\tan \beta)_{\max} + 3M + 18} \tag{2.15}$$

Chu & Lo's relation according to equation 2.15 is added in Figure 2.21 for different critical angles and compared with Bolton's relation.

In Figure 4.4 (page 55) several constitutive models are compared with Rowe's and Chu & Lo's relation.

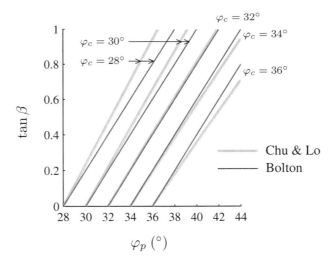

Figure 2.21: Bolton's relation according to equation 2.13 is compared with predictions by Chu & Lo according to equation 2.15. The relations are plotted for different critical friction angles φ_c. For $\tan\beta = 0$, φ_p equals φ_c.

2.2   Specifics of clay behaviour

2.2.1   Compression behaviour

Isotropic normal compression: Various mathematical formulations are used to describe the compression behaviour of soil. More details to compression behaviour of sand is given in Section 2.3.1 (page 36).

Commonly assumed relations for clay are:

- The loading (and unloading) curve is linear in the $\ln p$ - $(1 + e)$ plot. This is assumed by Roscoe & Burland [58].

- The loading (and unloading) curve is linear in the $\ln p$ - $\ln(1 + e)$ plot. This assumption is based on studies by Butterfield [6], see also Mašín [44].

In order to estimate the difference between the two relations, they are plotted in Figure 2.22 and compared with experimental data. The softer the clay is, the better is the $\ln(1 + e)$ vs. $\ln p$ approach, according to Butterfield [6]. For not very soft clays, the difference between the two relations is small.

In this thesis the linear relation between $\ln p$ and $\ln(1 + e)$ is adopted:

$$\ln(1 + e) = N - \lambda^* \ln \frac{p}{\sigma^*} \qquad (2.16)$$

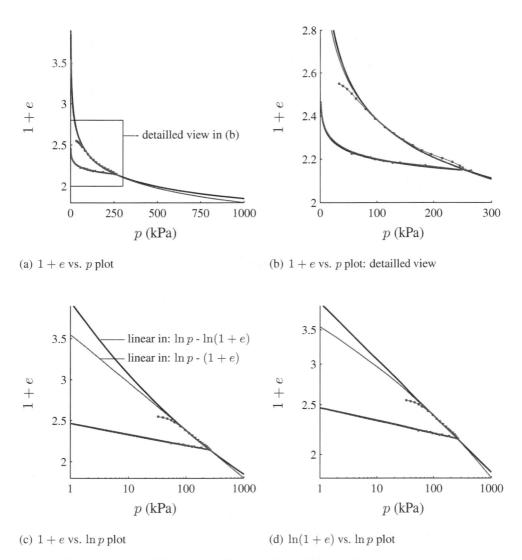

(a) $1 + e$ vs. p plot

(b) $1 + e$ vs. p plot: detailled view

(c) $1 + e$ vs. $\ln p$ plot

(d) $\ln(1 + e)$ vs. $\ln p$ plot

Figure 2.22: London clay: Comparison of the semi-logarithmic and log-log approaches to describe normal isotropic compression and unloading behaviour of clay.[24] Note that the difference of the representations is rather small for most clays, but becomes significant according to Butterfield [6] for more compressible clays.[25] The overconsolidated London clay (data from Mašín [44]) approaches both normal compression lines. The semi-logarithmic and log-log approaches to unloading are, both, appropriate for London clay.

N is the ordinate intercept, i.e. $N = \ln(1 + e)$ at $p = \sigma^* = 1$ kPa and λ^* is the slope[21] of the NCL, cf. Figure 2.23.

---

[21]Note that the quantities λ and κ are defined by a linear relation between $\ln p$ and $1 + e$, whereas the symbols λ^* and κ^* are used with reference to $\ln p$ - $\ln(1 + e)$ plots.

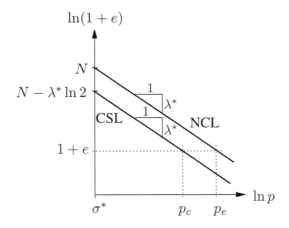

Figure 2.23: Normal Compression Line (NCL) and Critical State Line (CSL) for clay. Instead of *normal compression*, the term *virgin compression* could also be used.

Un-/Reloading under isotropic compression: The unloading stiffness is much higher than the stiffness at primary loading, see Figure 2.24(a). Ongoing reloading leads to an approach of the NCL. The complex soil behaviour is idealized in Figure 2.24(b). According to Critical State Soil Mechanics, the unloading stiffness is described by the parameter[23] κ^* in equation 2.17, cf. Figure 2.24:

$$\ln(1+e) = N - \lambda^* \ln \frac{p_0}{\sigma^*} + \kappa^* \ln \frac{p_0}{p} \qquad (2.17)$$

As for isotropic normal compression, linear relations for unloading can be found in the $\ln p$ - $(1+e)$ plot as well as in the $\ln p$ - $\ln(1+e)$ plot, cf. Figure 2.22. The difference between the two formulations for unloading is even smaller than at virgin loading.

2.2.2   Critical State

The critical state line (CSL, Figures 2.4(b) and 2.23) can be approximated as

$$\ln(1+e_c) = N - \lambda^* \ln \frac{2\,p}{\sigma^*} \qquad (2.18)$$

---

[24]Using the semi-logarithmic plot, the ordinate intercept is assumed to $N = 3.549$, the inclinations are $\lambda = 0.253$ and $\kappa = 0.058$; for the logarithmic formulation, the ordinate intercept is $N = 1.375$ and the inclinations are $\lambda^* = 0.11$ and $\kappa^* = 0.025$. The parameters except for κ and κ^* are chosen according to Mašín [43].

[25]Butterfield [6] mentions Mexico City clay, Newfoundland peat and Chicago clay, amongst others, as clays with higher compressibility. London clay is the soil with the highest compressibility among the clays investigated in this thesis (see Chapter 5).

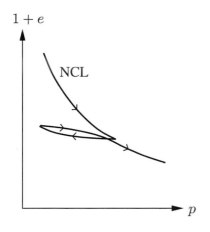

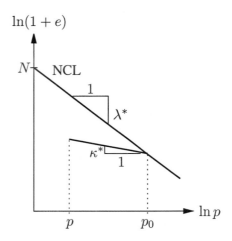

(a) Soil behaviour under hydrostatic loading, un- and reloading

(b) Primary loading is idealized through λ^*, un- and reloading stiffness is idealized through κ^*. Note that λ and κ is used in the $\ln p$ - $(1 + e)$ plot

Figure 2.24: Primary loading, unloading- and reloading stiffness

and is parallel to the NCL in the $\ln p$ - $\ln(1 + e)$ plot.[26] Note that according to the yield locus of the Modified Cam Clay Model introduced by Roscoe & Burland [58] the ordinate intercepts of the CSL and NCL are related, cf. Figure 2.25.

In the p - q plot critical states are characterized by the slope M, which is related to the critical friction angle φ_c, cf. equation 2.19 and Figure 2.4(a):

$$q = Mp \quad \text{with} \quad M = \frac{6 \sin \varphi_c}{3 - \sin \varphi_c} \tag{2.19}$$

2.2.3 State boundaries

Feasible states are limited with respect to stress $\mathbf{T}$ and void ratio e. E.g., the NCL is an upper bound for void ratios. The ratio of q to p for hydrostatic compression is zero. At critical states, this ratio is $q/p = 6 \sin \varphi_c / (3 - \sin \varphi_c)$ and e equals e_c.

In Figure 2.26(a) some directions of stretching are shown.[27] Assuming that soil cannot resist tensile stresses, the admissible stress states are limited[28] by the two lines

---

[26] As for the NCL, formulations for the CSL defined in the $\ln p$ - $\ln(1 + e)$ or the $\ln p$ - $(1 + e)$ plot can be found. The relation according to equation 2.18 is used by Mašín [44].

[27] Figure 2.26(a) is a Rendulic plot, i.e. axisymmetric states are considered.

[28] For $\sigma_1 \neq 0$ and $\sigma_3 = 0$ follows: $q = \sigma_1$ and $p = \sigma_1/3$ and therefore $q/p = 3$. On the extension side follows from $\sigma_1 = 0$ and $\sigma_3 \neq 0$, $q/p = -1.5$, respectively.

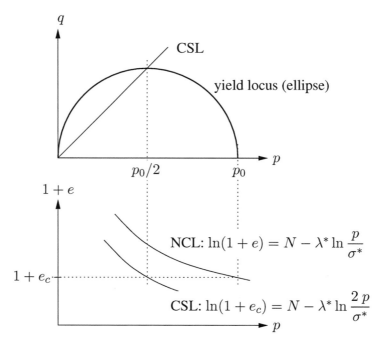

Figure 2.25: This Figure shows how the ordinate intercepts of the CSL and NCL are related according to the yield locus of the Modified Cam Clay model. Note that the formulations of the NCL and CSL in this plot differ from the original formulations in the MCC model.

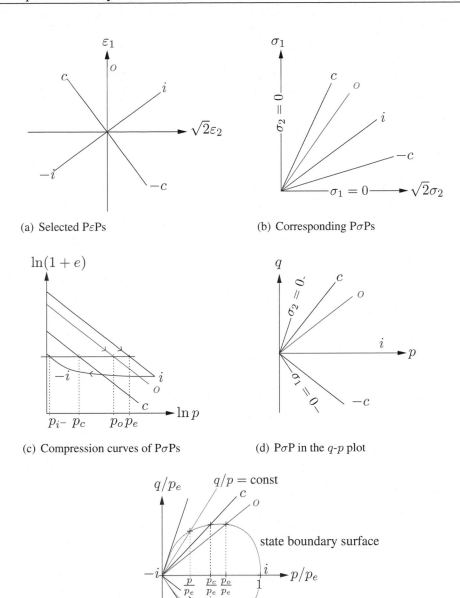

(a) Selected PεPs

(b) Corresponding PσPs

(c) Compression curves of PσPs

(d) PσP in the q-p plot

(e) From Figures 2.26(c) and 2.26(d) follows the shape of the state boundary surface.

Figure 2.26: Schematic description of state boundaries: i refers to isotropic compression, $-i$ to isotropic extension, c to isochoric triaxial compression and $-c$ to isochoric triaxial extension, o to oedometric compression

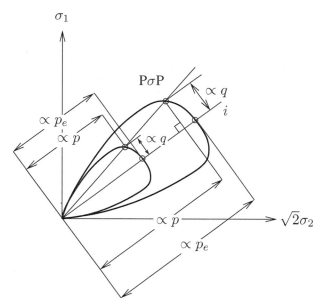

Figure 2.27: State boundary surfaces (schematically) for $e = $ const are limits for admissible stress states. Note that for a specific PσP for any void ratio, the ratio of p/p_e and q/p_e is the same.

$\sigma_1 = 0$ and $\sigma_2 = 0$, see Figure 2.26(b). Applying proportional strain paths to soil samples, proportional stress paths are reached with constant stretching $\mathbf{D} = $ const, i.e. when $\sigma_1 : \sigma_2 : \sigma_3 = $ const and $q/p = $ const, respectively, cf. Figures 2.12, 2.26(b) and 2.26(d). Figure 2.27 shows stress contour lines for a constant void ratio in the Rendulic plane. It is common to normalize stresses by the so-called *Hvorslev's equivalent consolidation pressure*:

$$p_e = \exp\left(\frac{N - \ln(1 + e)}{\lambda^*}\right) \qquad (2.20)$$

p_e is the value of mean stress on the isotropic normal consolidation line which refers to the current specific volume $(1 + e)$, cf. Figures 2.23 and 2.26(c). By analogy to Goldscheider's second rule, also in e-p plots curves obtained with initially virgin samples (so-called normal compression curves) are asymptotically approached by curves starting at oversonsolidated states. As the isotropic normal compression line, other normal compression lines are assumed linear in the $\ln p$ - $\ln(1 + e)$ plot. For a specific proportional path PσP, p/p_e is therefore constant and independent of the current void ratio e, as shown in Figure 2.27. The same applies for q/p_e. Proportional paths are therefore characterized by $p/p_e = $ const and $q/p = $ const, cf. Figures 2.26(d) and 2.26(c). From Figures 2.26(c) and 2.26(d) follows the so-called *state boundary surface* (SBS), see Figure 2.26(e). It proves useful to represent the SBS in q - p - e or q/p_e - p/p_e plots. All feasible states lie within the SBS, cf. Figure

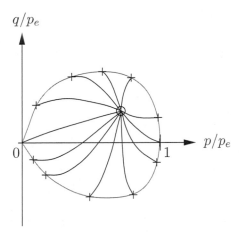

Figure 2.28: Schematically: Starting from state (o), the SBS is reached under constant stretching (+).

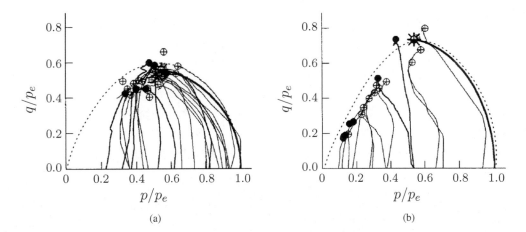

(a)

(b)

Figure 2.29: Normalized stress paths obtained with CU-tests of different layers of the subsoil of the tower of Pisa: (a) Layer A1, (b) Layer B1. The test data refer to CU-tests by Rampello & Callisto [57]. The (assumed) envelope of the normalized stress paths has the shape of the state boundary surface (dotted line).

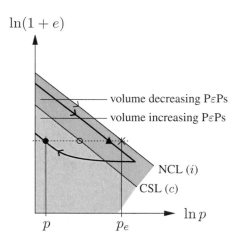

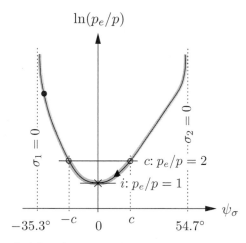

(a) After sufficiently long proportional stretching the memory is swept out and the distance from the isotropic normal compression line is defined through p_e/p =const, cf. Gudehus & Mašín [18] and Mašín [46].

(b) Gudehus & Mašín [18] propose this plot of p_e/p vs. ψ_σ, which is obtained by applying proportional stretching in the directions $-d < \psi_{\dot\varepsilon} < d$, according to Figure 2.13(a).

Figure 2.30: (a): Applying a volume decreasing PεP should result in a normal compression line between the NCL and CSL. Applying a volume increasing PεP should asymptotically lead to a line below the CSL (so-called *extension state*). A more general picture is given in (b): volume decreasing PεPs are characterized through paths with $-c < \psi_\sigma < c$. Volume increasing PεPs are characterized through paths with $-35.3° < \psi_\sigma < -c$ and $c < \psi_\sigma < 54.7°$. Note that according to Mašín [46] the *extension states* have not been confirmed by experiments, but they follow conceptually from many constitutive models. For better comprehension certain points ($\bullet$, $\circ$, $\times$, $\blacktriangle$) are marked in (a) and (b).

2.28. Under proportional stretching the memory is gradually swept out and the state boundary surface is reached, cf. Figure 2.28 and Mašín & Herle [48]. Figure 2.29 shows normalized stress paths obtained with CU-tests on Pisa clay. The (imagined) envelope of the stress paths is the state boundary surface.

Gudehus & Mašín [18] propose the following representation of admissible states with respect to void ratio and PσPs for clay. In Figure 2.30(b), it is proposed how PσPs (in terms of ψ_σ) are connected with p_e/p. The distance from the isotropic compression line is indicated by p_e/p. E.g. for hydrostatic compression applies $p_e/p = 1$ and $\psi_\sigma = 0$, at critical states p_e/p_c is assumed equal 2. Long enough proportional stretching will lead to constant values for p/p_e for states with $\delta \leq 0$ as well as for states $\delta > 0$, so-called *asymptotic extension states*, cf. Gudehus [17], Gudehus & Mašín [18], Mašín [46]. Figure 2.30(b) shows schematically how ψ_σ is related to p_e/p for any proportional path.

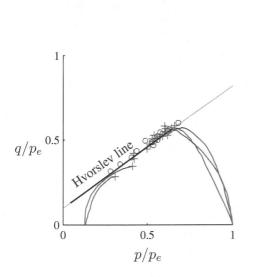

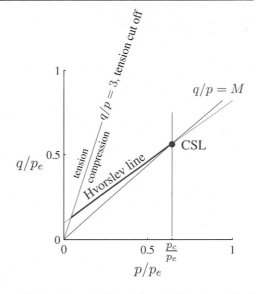

(a) Peaks of CU tests (+) and CD test (o) on Weald Clay lie on the Hvorslev line, Figure modified from Muir Wood [50] with data from Muir Wood [50] and Atkinson & Bransby [2];

(b) The Hvorslev line is bounded by the critical state point $q/p = M$, $p/p_e = p_c/p_e$ and by assuming only compression for soil: $q/p \leq 3$.

Figure 2.31: Hvorslev line

The Hvorslev line: Hvorslev analysed failure of overconsolidated Weald Clay and observed that peaks of drained and undrained tests lie on the same line, the so-called Hvorslev line, cf. Figure 2.31(a) and Atkinson & Bransby [2].[29] The Hvorslev line is bounded by the critical state point[30] $q/p_e/(p/p_e) = q/p = M$ and by assuming only compression for soil, the tension cut off is $q/p = 3$, cf. Figure 2.31(b).[31] For Weald Clay p_c/p_e is less than $2/3$ according to Muir Wood [50], see Figure 2.31(b). Note that according to the state boundary surface of the MCC model, p_c/p_e equals 2, cf. Figure 2.25.

---

[29]c_{pe} is often used as ordinate intercept of the Hvorslev line and $\tan \varphi_e$ is the inclination of the line in the q/p_e - p/p_e plot, cf. e.g. [50].

[30]Note that the CSL is a point in the q/p_e - p/p_e plot; p/p_e and $q/p = M$ are constants at critical states and therefore $q/p_e = \text{const}$;

[31]From $\sigma_1 = 0$ follows $q/p = 3$, from $\sigma_3 = 0$ follows $q/p = -1.5$.

2.3   Specifics of sand behaviour

2.3.1   Compression behaviour

As already shown in Figure 2.2, the compaction of sand under isotropic compression is much less compared with clay. In Figure 2.32 a schematic plot of compression curves of sand starting at e_{i0} and $e_{\min}$ is shown. e_{i0} denotes the maximum void ratio at zero stress level and $e_{\min}$ the minimum void ratio according to index tests. Note that void ratios at zero stress level cannot be determined experimentally, but there are meaningful estimates, cf. Section 2.3.4. Bauer [4] proposes the following relation for isotropic compression starting at e_{i0}:

$$e_i = e_{i0} \exp\left[-\left(\frac{3p}{h_s}\right)^n\right] \tag{2.21}$$

h_s, n are material parameters.

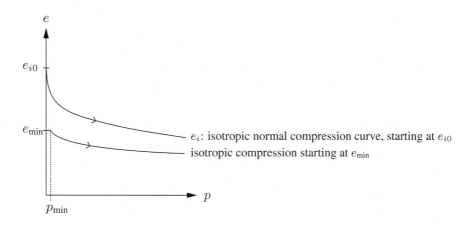

Figure 2.32: Schematic plot: Compression curves for sand starting at e_{i0}, the maximum void ratio at zero stress level and $e_{\min}$, i.e. the minimum void ratio at the mean pressure $p = p_{\min}$. Figure modified from Herle & Gudehus [23]

2.3.2   Critical state

Gudehus [16] proposes that the isotropic normal compression curve starting at e_{i0} is affine[32] to the critical state line e_c, cf. Figure 2.33.

$$\frac{e_c}{e_i} = \frac{e_{c0}}{e_{i0}} \tag{2.22}$$

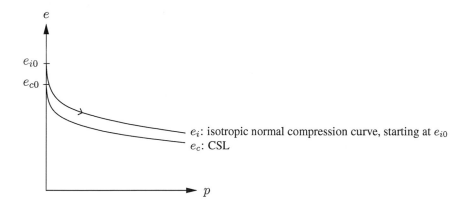

Figure 2.33: The CSL is assumed affine to the isotropic normal compression curve starting at e_{i0}, cf. Gudehus [16].

Gajo & Muir Wood [11, 12] assume the following relation for the CSL for sand in their model *Severn-Trent-Sand*, cf. Figure 2.34.

$$1 + e_c = v_\lambda - \lambda \ln \frac{p}{\sigma^*} \tag{2.23}$$

with the ordinate intercept $v_\lambda = 1 + e_c$ at the reference stress $\sigma^* = 1$ kPa.[33]

2.3.3 Minimum void ratio

The minimum void ratio e_d can be reached under cyclic shearing with small amplitudes starting from the compression curve shown in Figure 2.35 and in Herle & Gudehus [23]. It is assumed that $e_{\min}$ equals e_d at zero stres level, i.e. e_{d0}. According to Gudehus [16] the minimum void ratio e_d is as well affine to e_i:

$$\frac{e_d}{e_i} = \frac{e_{d0}}{e_{i0}} \tag{2.24}$$

2.3.4 Void ratio-pressure relation

Combining Figures 2.33 and 2.35 provides the assumption by Gudehus [16] that the isotropic normal compression curve e_i is affine to the critical state line e_c and to e_d, cf. Figure 2.36. The following equation was therefore introduced by Bauer [4]:

$$\frac{e_i}{e_{i0}} = \frac{e_c}{e_{c0}} = \frac{e_d}{e_{d0}} = \exp\left[-\left(\frac{3p}{h_s}\right)^n\right] \tag{2.25}$$

---

[32] Affine curves become parallel in a semilogarithmic plot.
[33] Note that here for sand the same formulation is used as in the Modified Cam Clay Model.

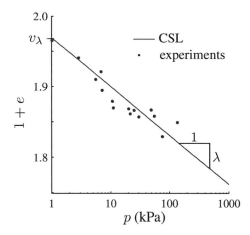

Figure 2.34: Critical states of Hostun Sand obtained by CU tests by various authors, data from Gajo & Muir Wood [11]. The CSL is assumed as a line in the semilogarithmic plot according to Gajo & Muir Wood [12].

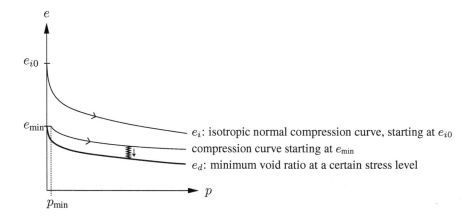

Figure 2.35: The minimum void ratio e_d at a certain stress level can be reached under cyclic shearing at constant pressure. Figure slightly modified from Herle & Gudehus [23]

Herle [22] and Herle & Gudehus [23] give a description of the calibration procedure. e_{d0} is approximately e_{min} and e_{c0} can be approximately set to e_{max}, obtained by index tests. It it advised to set e_{i0} equal to $1.2\ e_{max}$.

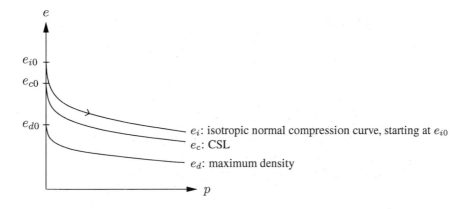

Figure 2.36: Gudehus [16] assumes that the isotropic normal compression curve e_i is affine to the critical state line e_c as well as to e_d-line.

Chapter 3

Some constitutive models for clay

In this section two models for clay, the Modified Cam Clay model and the hypoplastic model of Mašín, are presented. In the following sections, simulations with Barodesy are compared with these models.

3.1 Modified Cam Clay model

The Modified Cam Clay model (MCCM)[1] is an elastoplastic model with hardening, introduced by Roscoe & Burland [58]. Muir Wood [50] gives a comprehensible summary of the MCCM.

Elasto-plastic stiffness matrix: With the elasto-plastic stiffness matrix $\mathbf{D_{ep}}$ the stress rates $\dot{p}$ and $\dot{q}$ for plastic loading are given by:

$$\begin{pmatrix} \dot{p} \\ \dot{q} \end{pmatrix} = \mathbf{D_{ep}} \begin{pmatrix} \dot{\varepsilon}_{\text{vol}} \\ \dot{\varepsilon}_q \end{pmatrix} \tag{3.1}$$

The general form of $\mathbf{D_{ep}}$, including a hardening function H, is:

$$\mathbf{D_{ep}} = \mathbf{D_e} - \frac{\mathbf{D_e} \frac{\partial g}{\partial \boldsymbol{\sigma}} \frac{\partial f}{\partial \boldsymbol{\sigma}}^T \mathbf{D_e}}{\frac{\partial f}{\partial \boldsymbol{\sigma}}^T \mathbf{D_e} \frac{\partial g}{\partial \boldsymbol{\sigma}} + H} \tag{3.2}$$

with the elastic stiffness matrix[2] $\mathbf{D_e}$, the direction of plastic strain $\partial g / \partial \boldsymbol{\sigma}$ (cf. Figure 3.1), and the yield function f.

The yield locus $f = 0$ is an ellipse in the p - q - space and is described by the following equation

$$f = \frac{q^2}{M^2} - p(p_0 - p) = 0 \tag{3.3}$$

---

[1]The Modified Cam Clay model is a further development of the Cam Clay model. Often the name Cam Clay model is used also for the MCCM.

[2]For linear elasticity, the elastic stiffness matrix $\mathbf{D_e}$ reads $\begin{pmatrix} K & 0 \\ 0 & 3G \end{pmatrix}$ with the bulk modulus K and the shear modulus G. For the MCCM nonlinear elasticity is used, cf. $\mathbf{D_e}$ in equation 3.5.

with the preconsolidation pressure p_0 (cf. Figure 3.1) and the slope M of the CSL (cf. Figures 2.4(a) and 3.1). Associated flow is assumed, i.e. the plastic potential g coincides with f. The plastic strain rates are related with the plastic potenial by $\dot{\varepsilon}^{pl} = \mu \partial g / \partial \sigma$. It follows that the direction of plastic deformation, i.e. the flow-rule of the MCC-model, can be described by:[3]

$$\frac{\dot{\varepsilon}^{pl}_{vol}}{\dot{\varepsilon}^{pl}_q} = \frac{M^2 - (q/p)^2}{2(q/p)} \tag{3.4}$$

The MCCM includes the linear relation in the $\ln p$ - $(1+e)$ plot for isotropic compression as shown in Figure 2.24(b). With the definition of the shear modulus $G = \dfrac{1}{3}\dfrac{\partial q}{\partial \varepsilon_q}$ the elastic stiffness matrix reads:[4]

$$\mathbf{D_e} = \begin{pmatrix} (1+e)p/\kappa & 0 \\ 0 & 3G \end{pmatrix} \tag{3.5}$$

The hardening parameter H is:

$$H = p\frac{(1+e)p_0}{\lambda - \kappa}(2p - p_0) \tag{3.6}$$

In this thesis the NCL is assumed accoring to Butterfield [6], i.e. the NCL is assumed as linear in the $\ln p$ - $\ln(1+e)$ plot, cf. Figure 2.24(b). Introducing the $\ln p$ - $\ln(1+e)$ formulation in the MCC-model, results in the following elastic stiffness matrix:

$$\mathbf{D_e} = \begin{pmatrix} p/\kappa^* & 0 \\ 0 & 3G \end{pmatrix} \tag{3.7}$$

and the hardening parameter H from equation 3.6 changes to:

$$H = p\frac{p_0}{\lambda^* - \kappa^*}(2p - p_0) \tag{3.8}$$

Figure 3.1 shows basic concepts of the Modified Cam Clay Model. A normally consolidated triaxial test is illustrated schematically. The size of the initial yield locus is defined through p_0 in Figure 3.1(a). Under conventional triaxial compression with

---

[3]Equation 3.4 links stress in terms of q/p with dilatancy $\dot{\varepsilon}^{pl}_{vol}/\dot{\varepsilon}^{pl}_q$, and therefore can be seen as the stress-dilatancy relation of the MCCM. In Figure 4.4 (page 55) equation 3.4 is plotted in a $\psi_{\dot{\varepsilon}}$ - ψ_σ plot and compared with other stress-dilatancy relations.

[4]Note that the shear modulus G can be independent of the mean stress p. It follows that the elastic shear stiffness is constant. Alternatively for a constant Poisson's ratio ν, the shear modulus G is a function of the mean pressure p.

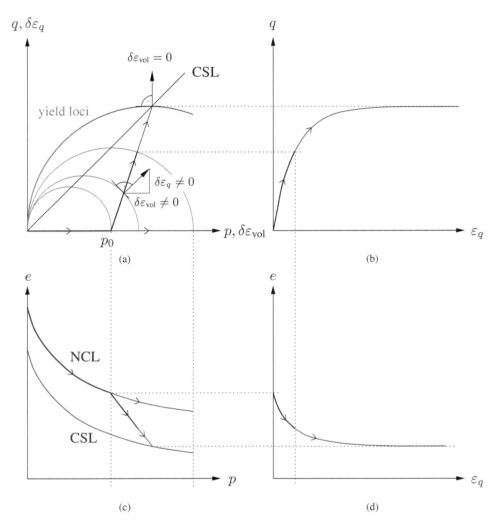

Figure 3.1: Graphical representation of the Modified Cam Clay Model (MCCM) according to Muir Wood [50]; Stress and strain paths of a **normally consolidated** triaxial test are illustrated schematically. The size of the initial yield locus is defined through p_0 in (a). Under conventional triaxial compression with $\dot{q}/\dot{p} = 3$, hardening occurs, i.e. the yield locus grows according to Figure (a). In Figure (b) the resulting q - ε_q relation is shown. In (c) the position of the e - p - curve related to the NCL and CSL is displayed. Figure (d) shows the e - ε_q relation, which coincides qualitatively with the ε_{vol} - ε_q relation.

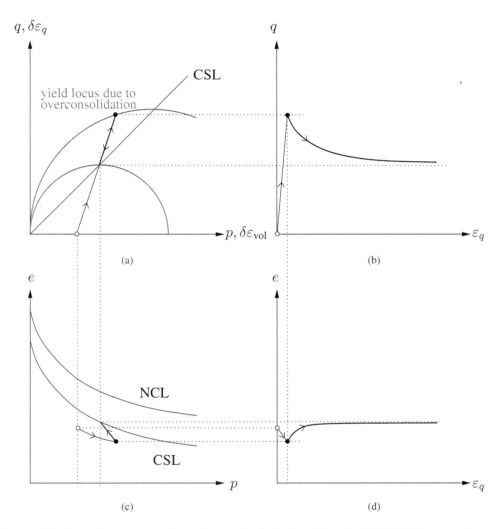

Figure 3.2: Graphical representation of the Modified Cam Clay Model (MCCM) according to Muir Wood [50]; Stress and strain paths of triaxial test with a **heavily overconsolidated** sample are illustrated schematically. The size of the initial yield locus is given through overconsolidation (a). Until the initial yield locus is reached, soil behaves elastic. Then softening occurs, i.e. the yield locus shrinks according to Figure (a) until the CSL is reached. In Figure (b) the resulting q - ε_q relation is shown. In (c) the position of the e - p - curve related to the NCL and CSL is displayed. Figure (d) shows the e - ε_q relation.

Parameter	Name	Calibration
φ_c	critical friction angle	CU-test: Figure 2.4(a)
N	ordinate intercept of the NCL	⎫ isotropic compression test:
λ, λ^*	slope of the NCL	⎬ Figures 2.23 and 2.24(b)
κ, κ^*	slope of the unloading line	⎭
G	shear modulus	initial stiffness at triaxial compression

Table 3.1: Material parameters of the MCC-model. The MCC-model originally refers to the parameters λ and κ. In this thesis the MCC-model is considered according to Butterfield, i.e. the NCL is assumed linear in the $\ln p$ - $\ln(1 + e)$ plot. The NCL is therefore described with the parameters λ^* and κ^*.

$\dot{q}/\dot{p} = 3$, hardening occurs, i.e. the yield locus grows according to Figure 3.1(a). Note that for a drained triaxial test of an overconsolidated sample, elastic behaviour is predicted until the yield locus is reached, cf. Figure 3.2. Then softening occurs, i.e. the yield locus shrinks according to Figure 3.2(a) until the CSL is reached.

Table 3.1 summarizes the material parameters of the MCC-model.

3.2 Hypoplastic model of Mašín

The hypoplastic model of Mašín [44] applies for clays and is based on other hypoplastic models (e.g. von Wolffersdorff [64], Herle & Kolymbas [24], Gudehus [16]) combined with ideas of Critical State Soil Mechanics: The compression relation and critical state concept according to Figures 2.23 and 2.24 have been introduced in this hypoplastic model.

The structure of the constitutive equation is:

$$\dot{\mathbf{T}} = f_s \mathcal{L}(\mathbf{T}) : \mathbf{D} + f_s f_d \mathbf{N}(\mathbf{T})|\mathbf{D}| \tag{3.9}$$

Hypoplastic models consist of a term linear in $\mathbf{D}$ and a nonlinear term. To describe the influence of the stress level (barotropy) and the influence of the density and the overconsolidation ratio (OCR) respectively (pyknotropy), the barotropy factor f_s and the pyknotropy factor f_d are used. The Mašín-model is summarized in Appendix B, its material parameters are summarized in Table 3.2.

Critical States in Hypoplasticity: With a tensorial function $\mathbf{B}$ defined according to Kolymbas [34]:

$$\mathbf{B} := \mathcal{L}^{-1} : \mathbf{N} \tag{3.10}$$

the constitutive equation reads:

$$\dot{\mathbf{T}} = f_s \mathcal{L} : (\mathbf{D} + f_d \mathbf{B}|\mathbf{D}|) \tag{3.11}$$

Parameter	Name	Calibration
φ_c	critical friction angle	CU-test: Figure 2.4(a)
N	ordinate intercept of the NCL	isotropic compression test:
λ^*	slope of the NCL	Figures 2.23 and 2.24(b)
κ^*	slope of the unloading line	
r	additional parameter	by parametric study

Table 3.2: Material parameters of the Mašín-model.

or:

$$\dot{\mathbf{T}} = f_s \mathcal{L} : (\mathbf{D}^0 + f_d \mathbf{B})|\mathbf{D}| \qquad (3.12)$$

To describe hypoplastic models in terms of elasto-plasticity, often the term *yield surface* is used. The shape of the yield surface is important for the simulation of critical state, that is $\dot{\mathbf{T}} = \mathbf{0}$ and $f_d = 1$ (in equation 3.12). This is obtained either with the trivial case $\mathbf{D} = \mathbf{0}$ or with $\mathbf{D}^0 = -\mathbf{B}$. Niemunis [53] calls $\mathbf{D}^0 = -\mathbf{B}$ the *flow rule*. Taking the norm we obtain:

$$|\mathbf{D}^0| = |\mathbf{B}| = 1 \qquad (3.13)$$

If $|\mathbf{B}| < 1$ the current stress state has not reached the yield surface so far. $|\mathbf{D}^0| = |\mathbf{B}|$ is fulfilled if $|\mathbf{B}| = 1$.[5] With the help of equation 3.13, a hypoplastic yield surface can be formulated as follows:

$$f(\mathbf{T}) = |\mathbf{B}| - 1 = 0 \qquad (3.14)$$

---

[5]This means that the actual stress has reached the critical state. $|\mathbf{D}^0|$ equals 1, by definition.

Chapter 4

Barodesy

The constitutive frame for soil called *barodesy*, proposed by Kolymbas [36, 37, 38, 39] and conceived as an evolution equation of the type $\mathring{\mathbf{T}} = \mathbf{h}(\mathbf{T}, \mathbf{D}, e)$, is based on the asymptotic behaviour of granulates. It can also be considered as a further development of Hypoplasticity [30, 31, 33, 35]. Barodesy is derived from the two rules by Goldscheider [14], cf. Section 2.1.3 on page 14. A detailled description of barodesy is presented in this Chapter and in Chapter 5. In the case of axial symmetry it is refered to Appendix C. The MATLAB source code is shown in Appendix D.

4.1 Mathematical formulation of Goldscheider's first rule

The directions $\mathbf{R}$ of proportional stress paths starting at $\mathbf{T} = \mathbf{0}$ and obtained with proportional strain paths $\mathbf{D} = \text{const}$ (Figures 2.12a and 2.12b) are given by a function $\mathbf{R}(\mathbf{D})$. In Kolymbas [36, 37, 38, 39], Medicus et al. [49] the following formulation of $\mathbf{R}$ is chosen:[1]

$$\mathbf{R} = \operatorname{tr} \mathbf{D}^0 \mathbf{1} + c_1 \exp(c_2 \mathbf{D}^0) \tag{4.1}$$

In this thesis the following alternative formulation is presented:

$$\mathbf{R} = -\exp(\alpha \mathbf{D}^0) \tag{4.2}$$

The determination of the scalar quantity α, is described in Section 4.4.1 (page 50). The differences between the two approaches are shown in Section 4.5.

---

[1] The exponential of a tensor $\mathbf{A}$ can be defined by means of its eigenvalues a_i:

$$\exp \mathbf{A} = \exp \begin{pmatrix} a_1 & 0 & 0 \\ 0 & a_2 & 0 \\ 0 & 0 & a_3 \end{pmatrix} = \begin{pmatrix} \exp a_1 & 0 & 0 \\ 0 & \exp a_2 & 0 \\ 0 & 0 & \exp a_3 \end{pmatrix}.$$

4.2   Mathematical formulation of Goldscheider's second rule

The asymptotic behaviour of proportional strain paths starting at $\mathbf{T} \neq \mathbf{0}$ as explained in Figure 4.1, implies:

$$\mathbf{T} + \dot{\mathbf{T}}\Delta t = \mu \mathbf{R}^0 \tag{4.3}$$

Rearranging equation 4.3 leads to $\dot{\mathbf{T}} = \hat{f}\mathbf{R}^0 + \hat{g}\mathbf{T}$, and with $\hat{f} := f \cdot \dot{\varepsilon} \cdot h(\sigma)$ and $\hat{g} := g \cdot \dot{\varepsilon} \cdot h(\sigma)/\sigma$ the general form of the barodetic constitutive relation is obtained (Kolymbas [38]):

$$\dot{\mathbf{T}} = h(\sigma) \cdot (f\mathbf{R}^0 + g\mathbf{T}^0) \cdot \dot{\varepsilon} \tag{4.4}$$

with

$$h = c_3 \sigma^* (\sigma/\sigma^*)^{c_4}, \tag{4.5}$$

where c_3 and c_4 are material constants.[2] The reference stress σ^* is set to 1 kPa. As stated in Section 1.3, the abbreviation σ is used for the Euclidean norm of $\mathbf{T}$, i.e. $\sigma = |\mathbf{T}|$ and $\dot{\varepsilon} = |\mathbf{D}|$. The scalar quantities f and g take into account critical states, the influence of stress level (*barotropy*) and density (*pyknotropy*).

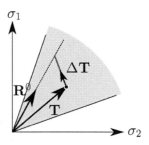

Figure 4.1: Mathematical realisation of Goldscheider's second rule: The stress increment $\Delta \mathbf{T}$ points to the corresponding proportional stress path $\mathbf{T} + \Delta \mathbf{T} = \mu \mathbf{R}^0, 0 \leq \mu \leq \infty$.

4.3   The scalar quantities f and g

For proportional paths $\mathbf{R}^0$ equals $\mathbf{T}^0$ and, hence, equation 4.4 yields:

$$\dot{\mathbf{T}} = h(\sigma) \cdot (f + g) \cdot \mathbf{T}^0 \cdot \dot{\varepsilon} \tag{4.6}$$

Critical states are characterized by vanishing stiffness (stress rate $\dot{\mathbf{T}} = \mathbf{0}$) and isochoric deformation $\delta = 0$. According to equation 4.4, the condition $\dot{\mathbf{T}} = \mathbf{0}$

---

[2]Instead of equation 4.5, $h = c_3 \sigma^{c_4}$ can be used. This implies that c_3 has the dimension (stress unit)$^{1-c_4}$, see also [37].

implies $\mathbf{R}^0 = \mathbf{T}^0$ and $f + g = 0$. One possible equation which fulfills $f + g = 0$ for critical states ($e = e_c$ and $\delta = 0$) is:[3]

$$f + g = \left(\frac{1+e}{1+e_c}\right)^{c_5} - 1 = 0 \tag{4.7}$$

The following generalisation for non-isochoric deformations ($\delta \neq 0$) is assumed:

$$f + g = \left(\frac{1+e}{1+e_c}\right)^{c_5} - 1 + \beta \cdot \delta \tag{4.8}$$

For proportional paths[4] we have $\dot{\mathbf{T}} = \dot{\sigma}\mathbf{T}^0$, thus we get from equation 4.6 and with h from equation 4.5:

$$\dot{\sigma} = c_3\sigma^* \left(\frac{\sigma}{\sigma^*}\right)^{c_4} \cdot (f + g) \cdot \dot{\varepsilon} \tag{4.9}$$

With f and g from equation 4.8 we have:

$$\dot{\sigma} = c_3\sigma^* \left(\frac{\sigma}{\sigma^*}\right)^{c_4} \cdot \left(\left(\frac{1+e}{1+e_c}\right)^{c_5} - 1 + \beta \cdot \delta\right) \cdot \dot{\varepsilon} \tag{4.10}$$

Requiring that $f + g$ is constant for PσPs, yields that $(1 + e)/(1 + e_c)$ is as well constant[5] for PσPs, see Chapter 5.2. We therefore may simplify equation 4.9 for proportional paths as

$$\frac{\dot{\sigma}}{\dot{\varepsilon}} \propto \left(\frac{\sigma}{\sigma^*}\right)^{c_4} \tag{4.11}$$

Comparing Janbu's or Ohde's relation in equation 2.2 and Figure 2.5(a) with equation 4.11, which follows from barodesy, yields $c_4 = 1$ for clays and $c_4 < 1$ for sands. In the subsequent sections it is shown how the constants c_3, c_4 and the scalar quantity β (which is not a constant, as it depends on δ) are determined. The scalar quantity β in equation 4.8 governs the loading- and unloading stiffness.

For general (i.e. non-proportional) paths, the scalar quantity $f + g$ needs to be splitted. The following parts are chosen:

$$f = c_6 \cdot \beta \cdot \delta - \frac{1}{2} \tag{4.12}$$

$$g = (1 - c_6) \cdot \beta \cdot \delta + \left(\frac{1+e}{1+e_c}\right)^{c_5} - \frac{1}{2} \tag{4.13}$$

---

[3]Other possibilities have been tried out as well, Medicus et al. [49] for example.

[4]For proportional compression paths the directions of $\dot{\mathbf{T}}$ and $\mathbf{T}^0$ coincide, i.e. $\mathbf{T}^0 = \dot{\mathbf{T}}/|\dot{\mathbf{T}}|$ is equal to $\mathbf{T}^0 = \mathbf{T}/|\mathbf{T}|$. From $\dot{\mathbf{T}} = \dot{\sigma}\mathbf{T}^0$, it follows that $\dot{\sigma}$ equals $|\dot{\mathbf{T}}| = \sqrt{\text{tr}\,\dot{\mathbf{T}}^2}$. Note that for extension paths $\dot{\sigma}$ equals $-|\dot{\mathbf{T}}|$.

[5]If $(1 + e)/(1 + e_c)$ is constant for PσPs, p/p_e is constant as well, see Figure 2.26(c).

4.4   The R - function: a rule for proportional paths

4.4.1   Determination of the scalar quantity α

In barodesy the directions of PσPs are given in dependence of the corresponding PεP by the **R** - function and are independent of the chosen scalar quantities f and g. The following procedure is proposed to fit the scalar quantity α in equation 4.2. As stated before, for proportional paths $\mathbf{R}^0$ equals $\mathbf{T}^0$.

With $K := \dfrac{\sigma_2}{\sigma_1} = \dfrac{R_2}{R_1} = \exp\left(\alpha D_2{}^0 - \alpha D_1{}^0\right)$, it follows:

$$\alpha = \frac{\ln K}{D_2{}^0 - D_1{}^0} \tag{4.14}$$

Replacing $D_1{}^0 - D_2{}^0$ with the scalar measure of deviatoric strain $\sqrt{3J_2}$ yields:

$$\alpha = \frac{\ln K}{\sqrt{3J_2}} \tag{4.15}$$

with $J_2 = 1/2 - \delta^2/6$ according to equation 1.9.

In order to find a function to describe several stress ratios K of proportional paths in dependence of stretching, it is suitable to choose the so-called strain increment ratio[6] m as a stretching invariant, cf. equations 2.8 and 2.9 in Section 2.1.4. Figures 2.16(c) on page 21 and Figure 4.2(a) show the relation proposed in equation 2.8 by Luo et al. [42]. Equation 2.8 captures states with $m > 0$ and peaks, in a certain range of $m < 0$.[7] Further decreasing of m results in decreasing of q/p, cf. Figure 4.2(a). Equation 2.8 is shown in a m - K plot in Figure 4.2(b).[8] The following relation (see Figures 4.2(a) and 4.2b) is proposed to approximate Chu & Lo's relation and to extend it for any proportional path.[9]:

$$K = 1 - \frac{1}{1 + c_1(m - c_2)^2} \tag{4.16}$$

c_1 and c_2 are constants. Equation 4.16 describes proportional paths for axisymmetric conditions. By including equations 4.15 and 4.16 into equation 4.2 a general criterion is obtained.

The following requirements are captured by equation 4.16:

---

[6] $m = -3\delta/\sqrt{6 - 2\delta^2}$, see also Section 1.3 (equations 1.8 and 1.10).

[7] The stress ratio q/p cannot increase unlimited, as tensile stresses are not feasible.

[8] q/p is expressed by K and vice versa with $\dfrac{q}{p} = \dfrac{3 - 3K}{1 + 2K}$ and $K = \dfrac{3 - q/p}{3 + 2q/p}$.

[9] It is described below how the constants c_1 and c_2 are determined. In Figure 4.2(c) equation 4.16 is plotted in the q/p-tan β-diagram.

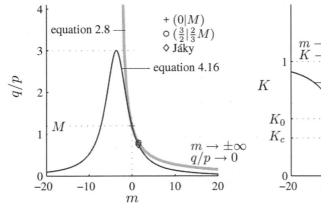

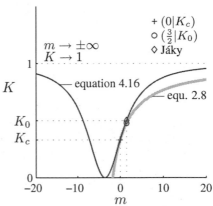

(a) The m - q/p relation (equation 4.16) captures states with $-\infty < m < \infty$. Limiting q/p to 3 prohibits tensile stresses.

(b) The m - K relation (equation 4.16) captures states with $-\infty < m < \infty$. Limiting K to 0 prohibits tensile stresses.

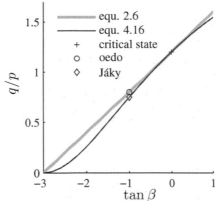

(c) Stress ratio q/p vs. dilatancy $\tan\beta$, according to Chu & Lo [8] (equation 2.6) is compared with equation 4.16.

Figure 4.2: Stress ratios q/p and K from equations 2.6, 2.8 and 4.16 are plotted vs. m and $\tan\beta$, respectively. For this plot φ_c is chosen to $30°$. For other critical friction angles the results are comparable.

- The stress ratio K must be positive, i.e. the mobilized friction angle must be $\leq 90°$.

- The stress ratio at critical state ($m = 0$) is $K_c = \dfrac{1 - \sin \varphi_c}{1 + \sin \varphi_c}$.

- The stress ratio for oedometric compression ($m = 3/2$) should match Jáky's relation:[10] $K_0 = 1 - \sin \varphi_c$.

- For isotropic compression and extension ($m \to \pm\infty$), K equals 1.

These requirements are fulfilled by equation 4.16 and result in the parameters c_1 and c_2 presented in Table 4.1. The directions of proportional paths are fixed by the parameter φ_c. Table 4.1 summarizes how the directions $\mathbf{R}(\mathbf{D})$ are determined.

$$\mathbf{R} = -\exp(\alpha \mathbf{D}^0)$$

$$\alpha = \frac{\ln K}{\sqrt{3/2 - \delta^2/2}}$$

$$K = 1 - \frac{1}{1 + c_1(m - c_2)^2} \quad \text{with} \quad m = \frac{-3\delta}{\sqrt{6 - 2\delta^2}}$$

$$c_2 = \frac{3\sqrt{K_c(1 - K_c)K_0(1 - K_0)} + 3K_c(1 - K_0)}{2(K_c - K_0)}$$

$$c_1 = \frac{K_c}{c_2^2(1 - K_c)}$$

$$\text{with } K_0 = 1 - \sin \varphi_c, \quad K_c = \frac{1 - \sin \varphi_c}{1 + \sin \varphi_c} \quad \text{and} \quad \delta = \operatorname{tr} \mathbf{D}^0$$

Table 4.1: Calibration of the $\mathbf{R}$ - function on the basis of φ_c. Note that for isotropic compression K equals 1 and therefore $\ln K = 0$. As the deviatoric strain under isotropic compression equals 0, it follows $\alpha = 0/0$. Applying de l'Hôpital's rule, we get $\alpha = \displaystyle\lim_{m \to \pm\infty} \frac{\ln K}{\sqrt{3J_2}} = 0$.

In Figure 4.3, the results of equation 4.2 are presented in the q-p plot as well as in the three-dimensional stress space. Figure 4.3(a) shows a plot in the principal stress space. For constant values of δ, equation 4.2 describes cones. Note that the directrix

---

[10]Alternatively, the oedometric predictions according to Chu & Lo [8] could be chosen, but the overall accordance is better with Jáky's formula.

of the cone resembles circles close to the isotropic axis, and becomes triangular at the edges of the compression octant.[11] Figure 4.3(b) shows cross sections of **R** - cones with the deviatoric plane. The cross sections refer to fixed δ values. Figure 4.3(c) shows a $\psi_{\dot{\varepsilon}}$ - ψ_{σ} plot according to Figure 2.13 (page 18) and Figure 4.3(d) shows the respective q/p - δ plot. Figures 4.3 (b) and (c) apply for axisymmetric conditions. For better comprehension, certain points are marked in the three plots.

In Figure 4.4 $\psi_{\dot{\varepsilon}}$ - ψ_{σ} plots are shown. Barodesy predictions are compared with predictions by clay hypoplasticity with explicitly defined asymptotic states by Mašín [47]. Note that barodesy (equation 4.2) gives proportional stress paths for all $\psi_{\dot{\varepsilon}}$, i.e. $-180° < \psi_{\dot{\varepsilon}} < 180°$. In contrast, the model by Mašín [47] predicts proportional stress paths for proportional strain paths in the range $-125.3° < \psi_{\dot{\varepsilon}} < 144.7°$, which corresponds to the limits proposed by Gudehus & Mašín [18]. Discrete element simulations by Mašín [46] showed that asymptotic states could only be obtained in a narrower range of $\psi_{\dot{\varepsilon}}$. The flow rule of the Modified Cam Clay model from equation 3.4 (page 42) is also plotted. For certain $\psi_{\dot{\varepsilon}}$ the MCC model predicts tensile stresses. For $\psi_{\dot{\varepsilon}} = -90°$, the MCC model overestimates the critical friction angle. In Figure 4.4(b) a detail of the $\psi_{\dot{\varepsilon}}$ - ψ_{σ} plot is shown and Chu & Lo's relation (equation 2.6, page 20) and Rowe's stress-dilatancy relation (equation 2.5, page 20) are added. Chu & Lo's relation for stress ratio vs. dilatancy (Section 2.1.4, page 20) is plotted for $0° < \psi_{\dot{\varepsilon}} < 115.2°$, which corresponds to $-3 < \tan\beta < 1.5$. It compares satisfactory with barodesy, see also Figure 4.5(a). Chu & Lo's relation only applies for contractant (i.e. $-3 < \tan\beta < 0$) and slightly dilatant (i.e. $\tan\beta \lesssim 1$) proportional paths. Rowe's relation predicts stress-dilatancy behaviour of CD-tests (including peak states). Barodesy can therefore only be compared with Rowe's relation for $\psi_{\dot{\varepsilon}} \geq 90°$, cf. Figure 4.4(b). All relations presented in Figure 4.4 are equaly valid for the critical state ($\psi_{\dot{\varepsilon}} = 90°$). In Figure 2.19 experimental results for K_0 are compared with Jáky's and Chu & Lo's predictions. As $K_0 = 1 - \sin\varphi_c$ is included in the **R** - function, barodesy predictions agree with Jáky's equation, see Figures 4.4(b) and 4.5(b). Note that Jáky's and Chu & Lo's K_0 predictions are similar.

4.4.2 Critical states in barodesy

The prediction of critical states is one of the core elements of barodesy. The **R** - function implies the shape of the critical stress surface. At critical states the stress rate vanishes, $\dot{\mathbf{T}} = 0$. Consequently **R** is proportional to stresses at critical state **T**, i.e. $\mathbf{R}^0 = \mathbf{T}^0$ (see equation 4.4). At critical states, the dilatancy vanishes, i.e. $\delta = 0$,

---

[11] The cone for tr $\mathbf{D}^0 = 0$ is almost identical with the Matsuoka-Nakai surface, see Section 4.4.2.

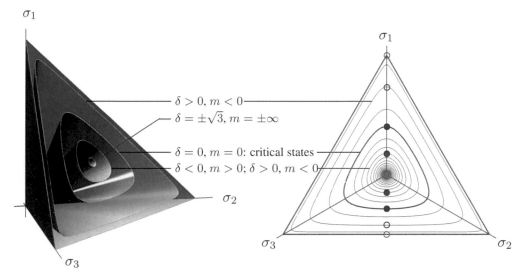

(a) 3D representation of surfaces formed by propor- (b) Deviatoric plane: Traces of proportional paths
tional paths with $\delta = $ const.   of prescribed dilatancy (constant δ - values)

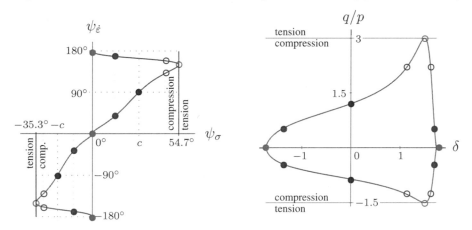

(c) $\psi_{\dot{\varepsilon}}$ - ψ_{σ} plot: ψ_{σ} is limited in the range from (d) q/p - δ plot: q/p is limited in the range from
$-35.3°$ to $54.7°$.   3 to -1.5.

Figure 4.3: PσPs obtained with equation 4.2: For the simulations $\varphi_c = 30°$ is chosen. Note that all
PσPs remain in the compression regime. For better comprehension, certain points are marked in (b),
(c) and (d).

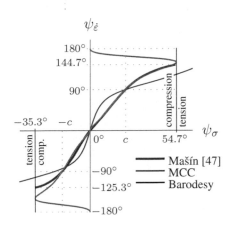

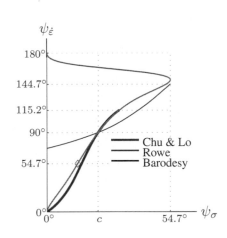

(a) $\psi_{\dot{\varepsilon}}$-ψ_σ plot obtained with equation 4.2, i.e. $\mathbf{R} = -\exp(\alpha\mathbf{D}^0)$. Barodesy is compared with predictions by clay hypoplasticity with explicitly defined asymptotic states by Mašín [47], and the Modified Cam Clay model. All models agree for critical state c, i.e. $\psi_{\dot{\varepsilon}} = 90°$. For $\psi_{\dot{\varepsilon}} = -90°$, the MCC model overestimates the critical friction angle. The MCC model predicts tensile stresses for certain $\psi_{\dot{\varepsilon}}$, i.e. $\psi_\sigma < -35.3°$ and $\psi_\sigma > 54.7°$.

(b) Detail of a $\psi_{\dot{\varepsilon}}$ - ψ_σ plot (Figure a): Chu & Lo's and Rowe's stress-dilatancy relations are compared with simulations with barodesy. Jáky's relation is marked (◇). Note that Rowe's relation predicts stress-dilatancy behaviour of CD tests (including peak states). The plots can therefore only be compared for $\psi_{\dot{\varepsilon}} \geq 90°$.

Figure 4.4: Stress-dilatancy relations are shown in $\psi_{\dot{\varepsilon}}$ - ψ_σ plots for London Clay, i.e. $\varphi_c = 22.6°$. Predicting PσPs with equation 4.2 provides results with compressive stresses, i.e. $-35.3° < \psi_\sigma < 54.7°$. The results according to barodesy are compared with clay hypoplasticity with explicitly defined asymptotic states by Mašín [47] and the MCC-model in (a) and Chu & Lo's relation and Rowe's stress-dilatancy relation for triaxial compression in (b).

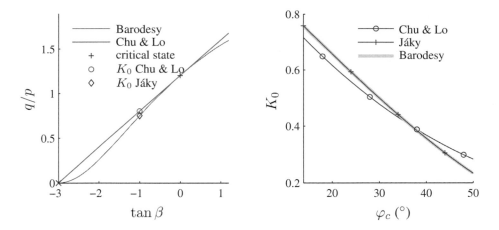

(a) Predictions with barodesy are in good agreement with Chu & Lo's predictions. φ_c is chosen to $22.6°$.

(b) K_0 in dependence of φ_c

Figure 4.5: Proportional paths with barodesy

and the asymptotic stress ratio is given by the critical friction angle φ_c, cf. Figure
2.4(a). From equations 4.2, 4.15 and 4.16 it follows (with $m = 0$):

$$\alpha = \sqrt{\frac{2}{3}} \ln K_c \qquad (4.17)$$

Locus of critical states in stress space

Kolymbas [39] describes how the locus of critical states can be determined with
equation 4.1. Eliminating $\mathbf{D}^0$ from equation 4.2, we get:

$$\mathbf{D}^0 = \frac{1}{\alpha} \ln(-\mathbf{R}) \qquad (4.18)$$

Taking into account that tr $\mathbf{D}^0$ equals 0, gives[12]

$$\frac{1}{\alpha} \ln(-R_1) + \frac{1}{\alpha} \ln(-R_2) + \frac{1}{\alpha} \ln(-R_3) = 0$$
$$\ln\left(-R_1 R_2 R_3\right) = 0$$
$$R_1 R_2 R_3 = -1 \qquad (4.19)$$

---

[12]Note that $R_i < 0$.

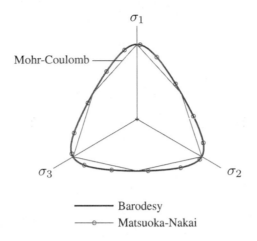

Figure 4.6: The critical state surface practically coincides with the Matsuoka-Nakai failure criterion. For comparison, Mohr-Coulomb's failure surface is added.

Equation 4.18 yields with $|\mathbf{D}^0| = 1$:

$$(\ln(-R_1))^2 + (\ln(-R_2))^2 + (\ln(-R_3))^2 = \alpha^2 \tag{4.20}$$

Considering $\boldsymbol{\sigma} = -\mu\mathbf{R}$ at critical states yields:[13]

$$\left(\ln\frac{-\mu R_1}{\mu}\right)^2 + \left(\ln\frac{-\mu R_2}{\mu}\right)^2 + \left(\ln\frac{-\mu R_3}{\mu}\right)^2 = \alpha^2$$

$$\left(\ln\frac{\sigma_1}{\mu}\right)^2 + \left(\ln\frac{\sigma_2}{\mu}\right)^2 + \left(\ln\frac{\sigma_3}{\mu}\right)^2 = \alpha^2$$

With equation 4.19[14] and α from equation 4.17 we get the locus of critical states in the stress space:

$$\left(\ln\frac{\sigma_1}{\sqrt[3]{\sigma_1\sigma_2\sigma_3}}\right)^2 + \left(\ln\frac{\sigma_2}{\sqrt[3]{\sigma_1\sigma_2\sigma_3}}\right)^2 + \left(\ln\frac{\sigma_3}{\sqrt[3]{\sigma_1\sigma_2\sigma_3}}\right)^2 = \frac{2}{3}\left(\ln\frac{1-\sin\varphi_c}{1+\sin\varphi_c}\right)^2 \tag{4.21}$$

Fellin & Ostermann [10] found that the critical state locus (the cone for $\delta = 0$ in Figure 4.3 a) practically coincides with the locus according to Matsuoka-Nakai, cf. Figure 4.6:

$$\frac{(\sigma_1 + \sigma_2 + \sigma_3)(\sigma_1\sigma_2 + \sigma_1\sigma_3 + \sigma_2\sigma_3)}{\sigma_1\sigma_2\sigma_3} = \frac{9 - \sin^2\varphi_c}{1 - \sin^2\varphi_c} \tag{4.22}$$

---

[13]In this case the more familiar symbol $\boldsymbol{\sigma} = -\mathbf{T}$ is used: $\boldsymbol{\sigma} = \begin{pmatrix} \sigma_1 & 0 & 0 \\ 0 & \sigma_2 & 0 \\ 0 & 0 & \sigma_3 \end{pmatrix}$.

[14]With $\boldsymbol{\sigma} = -\mu\mathbf{R}$ and $R_1 R_2 R_3 = -1$ follows: $\sigma_1\sigma_2\sigma_3 = \mu^3$ and $\mu = \sqrt[3]{\sigma_1\sigma_2\sigma_3}$

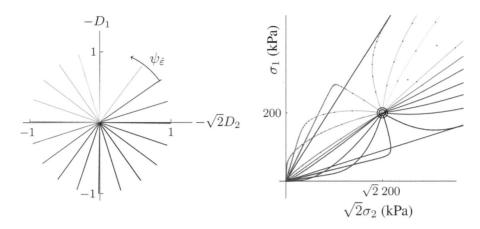

(a) Directions of PεPs characterized through $\psi_{\dot{\varepsilon}}$ (b) Approach to PσPs

Figure 4.7: Simulations with barodesy: Starting at $\mathbf{T} \neq \mathbf{0}$, proportional strain paths lead asymptotically to the corresponding proportional stress paths starting at $\mathbf{T} = \mathbf{0}$.

4.5   Proportional paths starting from $\mathbf{T} \neq 0$

Figure 4.7(a) shows the directions of proportional strain paths. In Figure 4.7(b) the corresponding stress paths (starting from $\mathbf{T} \neq \mathbf{0}$) are shown. Choosing $\mathbf{R}$ according to equation 4.2 and with f and g from equations 4.12 and 4.13, we get stress paths according to Figure 4.7(b).[15] Equation 4.1 yields the directions $\mathbf{R}$ in Figure 4.8(a) obtained by stretchings $\mathbf{D}$ according to Figure 4.7(a). Figure 4.8(b) shows the asymptotic approach of stress paths to the corresponding PσPs. Note that certain stress paths include tensile stresses. Figure 4.8(c) shows the ψ_{σ}-$\psi_{\dot{\varepsilon}}$ relation, obtained by equation 4.1. To avoid tensile stresses, $\mathbf{R}$ according to equation 4.2 was introduced in this thesis. As shown in Section 4.4 and in this Section, the new approach yields satisfactorily results.

Critical states in p vs. e plot

As described in Section 2.2.2 (page 28), critical states can also be indicated in the mean stress p - void ratio e plot as a relation between p and e_c. The pressure de-

---

[15]The actual stiffness is also visualized in Figure 4.8(b). The stress change to obtain a constant strain change (i.e. $\Delta \varepsilon = \sqrt{\dot{\varepsilon}_1^2 + \dot{\varepsilon}_2^2 + \dot{\varepsilon}_3^2} \Delta t$) is represented by the distance between two dots. The smaller this distance is, the smaller is the required stress change: If the dots lie close to each other, the soil is softer than if the distance of the dots is large.

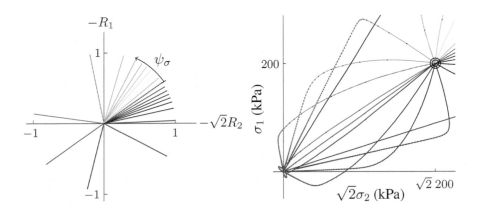

(a) The directions of $\mathbf{R}(\mathbf{D})$ are the directions of PσPs, characterized through ψ_σ.

(b) Stress paths starting from $\mathbf{T} \neq \mathbf{0}$ are attracted by the corresponding PσPs, cf. Figure (a).

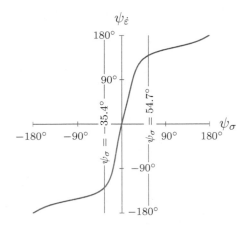

(c) The $\psi_{\dot\varepsilon}$-ψ_σ plot connects the directions of PεPs with the directions of PσPs. Certain PσPs lie in tensile areas, i.e. $\psi_\sigma < -35.3°$ and $\psi_\sigma > 54.7°$.

Figure 4.8: In this Figure $\mathbf{R}$ is chosen according to equation 4.1, i.e. $\mathbf{R} = \operatorname{tr}\mathbf{D}^0\mathbf{1} + c_1\exp(c_2\mathbf{D}^0)$. Certain PσPs lie in tensile areas. This shows that $\mathbf{R}$ according to equation 4.1 is inappropriate, if only compressive stress states are assumed for soil. $\mathbf{R}$ according to equation 4.2 is therefore introduced in this thesis. Note the differences between Figures 4.7(b) and 4.8(b) and between Figures 4.3(c) and 4.8(c).

pendent critical void ratio e_c is included in barodesy, cf. equation 4.13. Figure 4.9 illustrates the approach to the CSL in the stress - and as well in the p-e plot.[16]

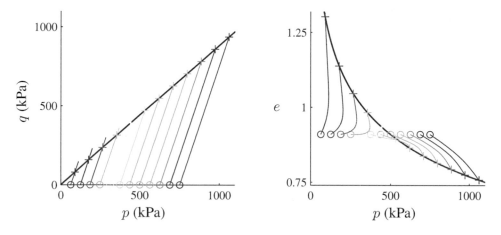

(a) Approach to the CSL in stress space   (b) Approach to the CSL in p-e space

Figure 4.9: The circles (o) are the starting points of simulations of drained triaxial tests with the same initial density. Critical states, marked with (+), are obtained when the CSL is reached in the stress (p-q) space as well as in the p-e space. Note the overconsolidated states at low p-values.

Determination of c_3

We consider the initial stiffness G of an undrained triaxial compression test. It is defined in equation 2.3 (page 11).

For an undrained test applies $\mathrm{tr}\,\mathbf{D} = 0$ (i.e. $\varepsilon_{\mathrm{vol}} = \varepsilon_1 + 2\varepsilon_2 = 0$) and therefore $\varepsilon_q = \frac{2}{3}(\varepsilon_1 - \varepsilon_2) = \varepsilon_1$. Hence, equation 2.3 reads for undrained deformation:

$$G = \frac{1}{3}\frac{\partial q}{\partial \varepsilon_1} = \frac{1}{3}\frac{\dot{T}_1 - \dot{T}_2}{D_1} \tag{4.23}$$

From equations 4.2 and 4.4 follows:

$$G = \frac{\dot{T}_1 - \dot{T}_2}{3D_1} = \frac{c_3\sigma^{c_4}\dot{\varepsilon}\left(fR_1^0 + gT_1^0 - fR_2^0 - gT_2^0\right)}{3D_1} \tag{4.24}$$

---

[16]Here are simulated CD-tests with clay. Similar simulations apply for sand.

The deformation in a CU-test starts from a hydrostatic stress state $\mathbf{T} = p_{\text{ini}}\mathbf{1}$ (i.e. $p_{\text{ini}} = T_1 = T_2 = T_3$). Introducing f from equation 4.12, $\mathbf{R}$ from equation 4.2, $\text{tr}\,\mathbf{D} = 0$, $\mathbf{D} = \mathbf{D}_{\text{cu}}$ and $G_0 = G|_{\delta_q = 0}$ into equation 4.24 yields:[17]

$$G = \frac{c_3 \sigma^{c_4} \dot{\varepsilon} \left(-1/2 R_1^0 + 1/2 R_2^0\right)}{3 D_1} = \frac{c_3 \sigma^{c_4} \sqrt{6} \left(R_1^0 - R_2^0\right)}{12} \qquad (4.25)$$

With $\mathbf{R}(\mathbf{D} = \mathbf{D}_{\text{cu}}) = \dfrac{1}{\sqrt{1 + 2K_c^2}} \begin{pmatrix} -1 & 0 & 0 \\ 0 & -K_c & 0 \\ 0 & 0 & -K_c \end{pmatrix}$ and $\sigma = \sqrt{3}p$, we obtain:

$$c_3 = \frac{12 G \sqrt{1 + 2K_c^2}}{\sqrt{6} \cdot (\sqrt{3}p)^{c_4} (K_c - 1)} \quad \text{with} \quad K_c = \frac{1 - \sin \varphi_c}{1 + \sin \varphi_c} \qquad (4.26)$$

4.5.1 Response envelopes

A stretching $\mathbf{D}$ indicated by the red line in Figure 4.10(a) causes a stress response according to Figure 4.11(a) where $\dot{\sigma}_1/\dot{\sigma}_2 < 1$.[18] Certain partitions of f and g cause pathological response envelopes, as shown for one example[19] in Figure 4.10(b).

Determination of c_6

In order to plot response envelopes according to Figure 4.11, it is proposed to calibrate c_6 as follows. Simplifying equation 4.4 for $e = e_c$ and $\mathbf{T}^0 = \mathbf{T}_{\text{iso}}^0 = -1/\sqrt{3}$ yields:

$$\dot{\mathbf{T}} = h(\sigma) \cdot (f \cdot \mathbf{R}^0 - g \cdot \mathbf{1}/\sqrt{3}) \cdot \dot{\varepsilon} \qquad (4.27)$$

Requiring $\dot{T}_1/\dot{T}_2 < 1$ for PσPs with $R_2 = 0$ and $R_1 < 0$, leads to:[20]

$$\frac{\dot{T}_1}{\dot{T}_2} = \frac{-f - g/\sqrt{3}}{-g/\sqrt{3}} < 1 \qquad (4.28)$$

$$f < 1 \qquad (4.29)$$

---

[17] Stretching for undrained triaxial compression yields $\mathbf{D}_{\text{cu}} = \begin{pmatrix} -1 & 0 & 0 \\ 0 & 0.5 & 0 \\ 0 & 0 & 0.5 \end{pmatrix} \cdot D_1$. It follows that $\dot{\varepsilon}$ equals $\sqrt{1 + 1/4 + 1/4} D_1$.

[18] Note that the red stress increment should point towards the $R_2 = 0$ - line, whereas the blue increment should point towards the $R_1 = 0$ - line.

[19] For the simulations in Figure 4.10(b), c_6 is chosen to 1.

[20] From $R_2 = 0$ and $R_1 < 0$ follows $\mathbf{R}^0 = \begin{pmatrix} -1 & 0 & 0 \\ 0 & 0 & 0 \\ 0 & 0 & 0 \end{pmatrix}$.

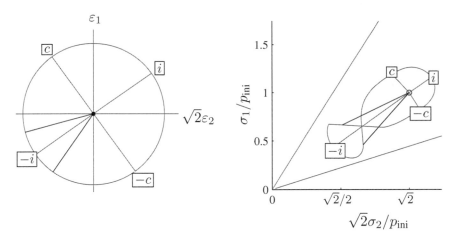

(a) Some directions of stretching **D**

(b) Graphical respresentation of response envelopes with the same relative density (i.e. $p_{ini}/p_e = $ const) normalized with the consolidation pressure p_{ini}.

Figure 4.10: A wrong partition of f and g causes pathological response envelopes.

With f from equation 4.12, we get:[21]

$$c_6 < \frac{1}{2\beta\delta} \tag{4.30}$$

With $\delta = \delta_{max} = \sqrt{3}$ we obtain:

$$c_6 = \frac{1}{2\beta(\delta = \sqrt{3})\sqrt{3}} \tag{4.31}$$

Note that the same shape of response envelopes is obtained with c_6 from equation 4.31, for arbitrary states $\mathbf{T} \neq \mathbf{T}_{iso}$ and $e \neq e_c$, cf. Figure 4.11(b).

It is described in Chapter 5, how the scalar quantity β and the constant c_4 are obtained to fit clay behaviour.

4.6 Mathematical discussion of barodesy

Barodesy (equation 4.4) predicts rate independent behaviour, as the constitutive equation is positive homogeneous of degree one in **D**:

$$\dot{\mathbf{T}}(\mathbf{T}, \lambda\mathbf{D}, e) = \lambda\dot{\mathbf{T}}(\mathbf{T}, \mathbf{D}, e)$$

---

[21]The scalar quantity β depends on δ. In Section 5.9 on page 66 is described how β is determined for clay. In order to determine c_6 in equation 4.31, we calculate β for $\delta = \sqrt{3}$.

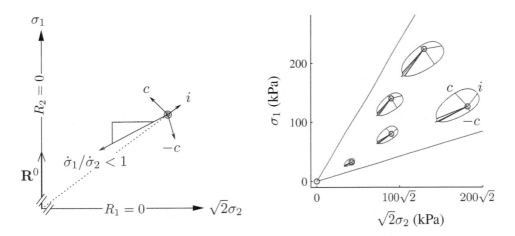

(a) It is demanded that the stress paths $\dot{\sigma}_1/\dot{\sigma}_2 < 1$ obtained by stretching marked by the red line in Figure 4.10(a) point towards $R_2 = 0$.

(b) The shape of response envelopes applies for arbitrary states with $\mathbf{T} \neq \mathbf{T}_{\text{iso}}$ and $e \neq e_c$.

Figure 4.11: Response envelopes with c_6 from equation 4.31

λ is a scalar quantity.

However equation 4.4 is inhomogeneous in $\mathbf{T}$:

$$\dot{\mathbf{T}}(\lambda\mathbf{T}, \mathbf{D}) \neq \lambda^n \dot{\mathbf{T}}(\mathbf{T}, \mathbf{D})$$

Changing the direction of loading (e.g. changing from $\mathbf{D}$ to $-\mathbf{D}$) leads to anelastic behaviour, i.e. different stiffnesses for loading and unloading. Equation 4.4 gives:

$$\dot{\mathbf{T}} = h(\sigma) \cdot f \cdot \mathbf{R}^0 \cdot |\mathbf{D}| + h(\sigma) \cdot g \cdot \mathbf{T}^0 \cdot |\mathbf{D}|$$

Changing from $\mathbf{D}$ to $-\mathbf{D}$ has no influence on $|\mathbf{D}|$, but on $\delta := \operatorname{tr}\mathbf{D}^0$, which is included in f and g. Therefore $\dot{\mathbf{T}}$ can be written as $\dot{\mathbf{T}} = F(|\mathbf{D}|) + G(\delta, |\mathbf{D}|)$. Starting at $\mathbf{T} \neq \mathbf{0}$ and applying a strain increment with $\delta < 0$, ($\delta > 0$ respectively) results in different stiffnesses, cf. Figure 4.12.

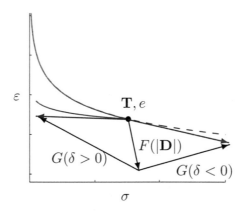

Figure 4.12: Different stiffnesses for loading and unloading are obtained with barodesy.

Chapter 5

Barodesy for clay

General properties of barodesy are presented in Chapter 4. The determination of the constant c_4 and the scalar quantity β is still missing. In this Chapter c_4 and β are chosen to fit clay behaviour. Common concepts of Critical State Soil Mechanics for clay are comprised.

5.1 Mathematical Formulation

Determination of c_4 and β: The normal compression line (equation 2.16) is used for the calibration of c_4 and β to achieve compliance with the NCL.[1]

Derivation of equation 2.16 with respect to time t gives:

$$\frac{\dot{e}}{1+e} = -\lambda^* \frac{\dot{p}}{p} \tag{5.1}$$

With $\frac{\dot{e}}{1+e} = \operatorname{tr} \mathbf{D} = -\sqrt{3}|\mathbf{D}| = -\sqrt{3}\dot{\varepsilon}$ and $\sigma = \sqrt{3}p$ the NCL reads[2]:

$$\dot{p} = \sigma \frac{1}{\lambda^*} \dot{\varepsilon} \tag{5.2}$$

With the general form of the barodetic constitutive relation, equation 4.4, isotropic compression (i.e. $\mathbf{T} = -p\mathbf{1}, \mathbf{R}^0 = \mathbf{T}^0 = -\frac{1}{\sqrt{3}}\mathbf{1}$) obtains the following form:

$$\dot{p} = c_3 \sigma^{c_4} \frac{f+g}{\sqrt{3}} \dot{\varepsilon} \tag{5.3}$$

Comparing equations 5.2 and 5.3 yields

$$c_4 = 1 \tag{5.4}$$

---

[1]Cf. similar approaches by Mašín [44, 45] and Medicus et al. [49]. In Medicus et al. [49] the procedure is the same, but the choice of f and g is different.

[2]The equation $\frac{\dot{e}}{1+e} = \operatorname{tr} \mathbf{D}$ holds for incompressible grains.

and

$$\frac{c_3}{\sqrt{3}}(f + g) = \frac{1}{\lambda^*} . \tag{5.5}$$

Now we write equation 4.8 for hydrostatic compression ($\delta = -\sqrt{3}$), use equation 5.5 and obtain

$$c_3 \left(\left(\frac{1+e}{1+e_c} \right)^{c_5} - 1 - \sqrt{3}\beta \right) = \frac{\sqrt{3}}{\lambda^*} \tag{5.6}$$

Introducing equations 2.16 and 2.18 into equation 5.6 leads to

$$\left(\frac{\exp(N - \lambda^* \ln(p/\sigma^*))}{\exp(N - \lambda^* \ln(2p/\sigma^*))} \right)^{c_5} - 1 - \sqrt{3}\beta = \frac{\sqrt{3}}{c_3 \lambda^*} \tag{5.7}$$

$$(\exp(N - \lambda^* \ln p - N + \lambda^* \ln 2 + \lambda^* \ln p))^{c_5} - 1 - \sqrt{3}\beta = \frac{\sqrt{3}}{c_3 \lambda^*}$$

$$\exp(c_5 \lambda^* \ln 2) - 1 - \sqrt{3}\beta = \frac{\sqrt{3}}{c_3 \lambda^*}$$

$$2^{c_5 \lambda^*} - 1 - \sqrt{3}\beta = \frac{\sqrt{3}}{c_3 \lambda^*}$$

$$\rightsquigarrow \beta = -\frac{1}{c_3 \lambda^*} + \frac{1}{\sqrt{3}} 2^{c_5 \lambda^*} - \frac{1}{\sqrt{3}} \tag{5.8}$$

Setting

$$\beta = -\frac{1}{c_3 \Lambda} + \frac{1}{\sqrt{3}} 2^{c_5 \lambda^*} - \frac{1}{\sqrt{3}} \tag{5.9}$$

with

$$\Lambda := -\frac{\lambda^* - \kappa^*}{2\sqrt{3}} \delta + \frac{\lambda^* + \kappa^*}{2} \tag{5.10}$$

yields the following:[3] Equation 5.5 is satisfied under isotropic compression, i.e. for $\delta = -\sqrt{3}$. Equations 5.9 and 5.10 ensure that a simulation of hydrostatic normal compression with barodesy starting from $e = \exp N - 1$ leads to the NCL, cf. Figure 5.1(a).

For unloading, derivation of equation 2.17 with respect to time t yields:

$$\frac{\dot{e}}{1 + e} = -\kappa^* \frac{\dot{p}}{p} \tag{5.11}$$

---

[3]Note that Λ in equation 5.10 is chosen as a function of δ. Λ equals λ^* for isotropic compression ($\delta = -\sqrt{3}$). For isotropic extension ($\delta = \sqrt{3}$), Λ equals κ^*. The consequences are described below. The values in between $\Lambda = \lambda^*$ and $\Lambda = \kappa^*$ are interpolated linearly.

With $\frac{\dot{e}}{1+e} = \operatorname{tr}\mathbf{D} = \sqrt{3}|\mathbf{D}| = \sqrt{3}\dot{\varepsilon}$ and $\sigma = \sqrt{3}p$ the unloading line reads:

$$\dot{p} = -\sigma \frac{1}{\kappa^*} \dot{\varepsilon} \tag{5.12}$$

Comparing equations 5.12 and 5.3 yields

$$\frac{c_3}{\sqrt{3}}(f+g) = -\frac{1}{\kappa^*} . \tag{5.13}$$

Substituting β for $\delta = \sqrt{3}$ from equation 5.9

$$\beta = -\frac{1}{c_3\kappa^*} + \frac{1}{\sqrt{3}}2^{c_5\lambda^*} - \frac{1}{\sqrt{3}} \tag{5.14}$$

into equation 5.13 with $\delta = \sqrt{3}$, we get

$$\left(\frac{1+e}{1+e_c}\right)^{c_5} - 1 + \sqrt{3}\beta = -\frac{\sqrt{3}}{\kappa^* c_3}$$

$$\left(\frac{1+e}{1+e_c}\right)^{c_5} - 1 - \frac{\sqrt{3}}{c_3\kappa^*} + 2^{c_5\lambda^*} - 1 = -\frac{\sqrt{3}}{\kappa^* c_3}$$

$$\left(\frac{1+e}{1+e_c}\right)^{c_5} = 2 - 2^{c_5\lambda^*}$$

$$\frac{1+e}{1+e_c} = \sqrt[c_5]{2 - 2^{c_5\lambda^*}} \tag{5.15}$$

The term $(1+e)/(1+e_c) = \text{const}$ in equation 5.15 indicates a line in the $\ln p$-$\ln(1+e)$ plot, cf. Figure 5.1(b). At this distance from the NCL the tangential unloading stiffness $\dot{p}/\dot{e}$ equals $-p/\kappa^*(1+e)$. Closer to the NCL the unloading stiffness is overpredicted compared with equation 5.11. For lower mean stresses p, the stiffness is underpredicted compared to equation 5.11.

Barodesy is summarized in Table 5.1.

5.2   Calibration

In Section 4.5 (page 60) it is proposed to calibrate c_3 by means of the initial stiffness G. In general, this procedure is helpful, but in some cases it may result in wrong predictions of extension states.[4] In Figure 5.2(a) simulations of isotropic compression tests are shown. Calibrating c_3 for a stiff sample, underpredicts the stiffness for

---

[4]c_3 does not only influence the initial stiffness, but also peak strength, unloading paths, etc.

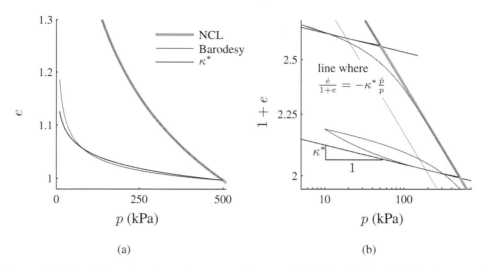

(a)   (b)

Figure 5.1: Simulation of an isotropic compression test: At the line indicated in (b) the tangential unloading stiffness $\dot{p}/\dot{e}$ equals $-p/\kappa^*(1+e)$. Above the line, the stiffness is underpredicted compared to equation 5.11.

unloading paths. Setting a lower limit for c_3 improves the predictions for isotropic extension, cf. Figure 5.2(a).[5]

The following procedure to calibrate c_3 is proposed:

With sufficiently long proportional stretching $p/p_e = p_i/p_e$ is constant for (isotropic) extension paths, cf. Section 2.2.3 and Figure 2.26(c) on page 31. It follows that the unloading stiffness in the $\ln p$ - $\ln(1+e)$ plot is characterized by the parameter λ^*, cf. Figure 5.2. From equations 5.1 and 5.3 and with $\mathrm{tr}\,\mathbf{D}^0 = \sqrt{3}$ follows for a proportional isotropic extension path:

$$c_3(f+g) = -\sqrt{3}/\lambda^* \tag{5.16}$$

With f, g and β from equations 4.12, 4.13 and 5.9 and with $(1+e)/(1+e_c) = (2 \cdot p_i/p_e)^{\lambda^*}$, we get:[6]

$$c_3\left(2\frac{p_i}{p_e}\right)^{\lambda^* c_5} - c_3 + c_3\sqrt{3}\left(-\frac{1}{c_3\kappa^*} + \frac{1}{\sqrt{3}}2^{c_5\lambda^*} - \frac{1}{\sqrt{3}}\right) = -\sqrt{3}/\lambda^* \tag{5.17}$$

Eliminating c_3 yields:

$$c_3 = \frac{-\sqrt{3}/\lambda^* + \sqrt{3}/\kappa^*}{2^{c_5\lambda^*} + \left(2\frac{p_i}{p_e}\right)^{c_5\lambda^*} - 2} \tag{5.18}$$

---

[5]The initial shear stiffness is then slightly underestimated, but the overall accordance is better.

[6]As shown in Figure 5.2(b), at isotropic extension $\ln(1+e)$ equals $N - \lambda^* \ln(p_e/p_i \cdot p/\sigma^*)$. We therefore get $\dfrac{1+e}{1+e_c} = \dfrac{\exp(N - \lambda^* \ln(p_e/p_i \cdot p/\sigma^*))}{\exp(N - \lambda^* \ln(2 \cdot p/\sigma^*))} = \left(2\dfrac{p_i}{p_e}\right)^{\lambda^*}$.

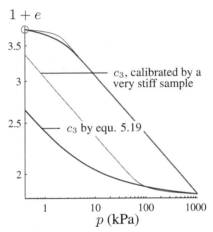
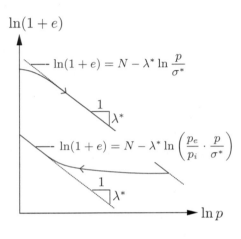

(a) Calibrating c_3 by a stiff sample, under-predicts the stiffness for unloading paths. Setting a lower limit for c_3 by equ. 5.19 improves the predictions for isotropic extension: upheavals are reduced by increasing the unloading stiffness.

(b) An isotropic compression and extension path

Figure 5.2: Isotropic compression is characterized by the NCL, isotropic extension is characterized by $N - \lambda^* \ln \left(p_e/p_i \cdot p/\sigma^* \right)$. Proportional compression and extension paths are characterized through $p/p_e = \text{const}$ and therefore $\dot{e}/(1+e) = -\lambda^* \cdot \dot{p}/p$.

$$\dot{\mathbf{T}} = c_3 \sigma^* (\sigma/\sigma^*)^{c_4} \cdot (f\mathbf{R}^0 + g\mathbf{T}^0) \cdot \dot{\varepsilon}$$
$$\mathbf{R} = -\exp(\alpha \mathbf{D}^0)$$

A

$$\alpha = \frac{\ln K}{\sqrt{3/2 - \delta^2/2}}$$

$$K = 1 - \frac{1}{1 + c_1(m - c_2)^2} \quad \text{with} \quad m = \frac{-3\delta}{\sqrt{6 - 2\delta^2}}$$

B

$$f = c_6 \cdot \beta \cdot \delta - \frac{1}{2}$$
$$g = (1 - c_6) \cdot \beta \cdot \delta + \left(\frac{1+e}{1+e_c}\right)^{c_5} - \frac{1}{2}$$

C

$$e_c = \exp\left(N - \lambda^* \ln \frac{2\,p}{\sigma^*}\right) - 1$$
$$\beta = -\frac{1}{c_3\Lambda} + \frac{1}{\sqrt{3}} 2^{c_5\lambda^*} - \frac{1}{\sqrt{3}}$$
$$\Lambda = -\frac{\lambda^* - \kappa^*}{2\sqrt{3}} \delta + \frac{\lambda^* + \kappa^*}{2}$$

Table 5.1: Equations A and B apply for sand and clay. Equations C apply only for clay. The reference stress σ^* equals 1 kPa, c_1 - c_6, λ^*, κ^* and N are material constants, cf. Table 5.2. The following abbreviations are used: $\delta = \operatorname{tr}\mathbf{D}^0, \dot{\varepsilon} = |\mathbf{D}|, \sigma = |\mathbf{T}|$. The superscript 0 marks a normalised tensor, i.e. $\mathbf{X}^0 = \mathbf{X}/|\mathbf{X}|$. In the case of axial symmetry it is referred to Appendix C. The MATLAB source code is shown in Appendix D in Listings D.2 and D.3.

Chosing $p_i/p_e = 1/100$ in equation 5.18 gives a lower limit for c_3, see Table 5.2 and equation 5.19:

$$c_3 \geq \frac{-\sqrt{3}/\lambda^* + \sqrt{3}/\kappa^*}{2^{c_5\lambda^*} + \left(2\frac{1}{100}\right)^{c_5\lambda^*} - 2} \tag{5.19}$$

Figure 5.3 shows how c_3 influences the ψ_σ - p_e/p plot. In Figure 2.30 on page 34 it is described how ψ_σ and p_e/p are related. Note that in Figure 5.3(a) p_e/p is underestimated for certain extension states. The consequences are seen at the unloading behaviour of isotropic compression tests ($\psi_\sigma = 0°$), cf. Figure 5.4. With relatively low values of p_e/p follows that rebound is overestimated, cf. Figure 5.4(a). Setting $p_e/p = 100$ results in the unloading behaviour in Figure 5.4(b). A more

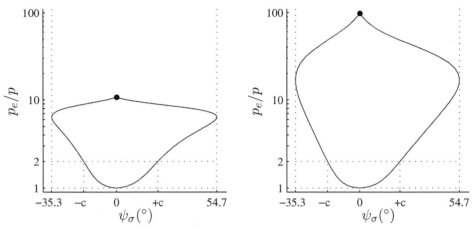

(a) Calibrating c_3 with a very stiff sample, under-
estimates certain values for p_e/p. Isotropic extension due to Figure 5.4(a) is marked (•).

(b) Setting $p_e/p = 100$ for isotropic extension
leads to this p_e/p - ψ_σ plot. Isotropic extension due to Figure 5.4(b) is marked (•).

Figure 5.3: Relation between p_e/p and ψ_σ, according to Figure 2.30

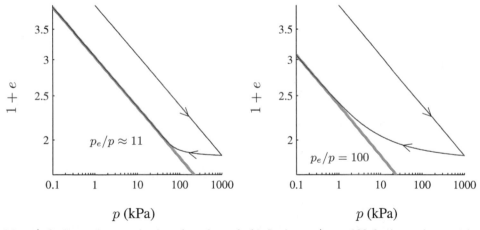

(a) p_e/p for isotropic extension is underestimated.
It follows that upheaval due to unloading is
overestimated.

(b) Setting $p_e/p = 100$ for isotropic extension
leads to the unloading behaviour in this plot.

Figure 5.4: Simulations of isotropic compression tests. The calibration of c_3 influences the unloading
behaviour. Isotropic extension is characterized through $\psi_\sigma = 0°$, cf. Figure 5.3.

general view is given in Figure 5.3. The tests from Figure 5.4 are marked in Figure
5.3. In Figure 5.5 state boundary surfaces are shown with different values for c_3.

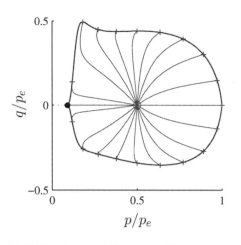 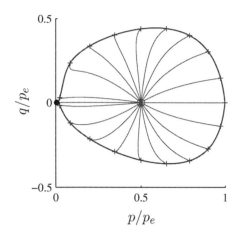

(a) Calibrating c_3 with a very stiff sample, may result in a qualitatively wrong shape of the SBS. Isotropic extension due to Figure 5.4(a) is marked ($\bullet$).

(b) Requiring $p/p_e \leq 1/100$ for isotropic extension yields a qualitatively correct shape of the SBS. Isotropic extension due to Figure 5.4(b) is marked ($\bullet$).

Figure 5.5: Calibration of c_3; the simulations refer to London clay. Note that the shapes qualitatively apply for other clays as well.

In order to calibrate the material constants c_1 - c_6 (Table 5.2), the parameters of Critical State Soil Mechanics (N, λ^*, κ^*, φ_c and G), which can be determined in an isotropic compression test and an undrained triaxial test, are sufficient. Both tests are parts of a consolidated undrained (CU) triaxial test. The slope λ^* and the intercept N of the NCL are obtained from isotropic compression, cf. Figures 2.23 and 5.9(b). The critical friction angle φ_c and the initial stiffness G can be determined by undrained triaxial compression tests, cf. Figures 2.4(a) and 2.7.

Remark: At very low stress levels, the unloading stiffness is underestimated, cf. the isotropic compression test in Figure 5.6. For certain loading/unloading ranges, upheavals are obtained. An easy way to increase the unloading stiffness is to decrease the parameter κ^*, e.g. by the factor $1/2$, cf. Figure 5.7. It follows that the stiffness close to the NCL is overestimated. Upheavals are reduced and the overall performance of the model is not very much affected. Another possibility is to set c_5 equal to 1, instead of $c_5 = 1/K_c$. $c_5 = 1/K_c$ increases the unloading stiffness, but reduces the overall performance of the model, e.g. the radial stress for oedo-

---

[7]$\hat{G}$ is calibrated by the author, the other parameters are from sources mentioned above.

*calibrated with experiment *PhM19* from [44];

[†]calibrated with experiments *CU 05, CU 04, CU 03*;

[8]San Francisco Bay Mud is only used to simulate critical strength; therefore only φ_c is calibrated.

$$c_1 = \frac{K_c}{c_2^2(1 - K_c)}$$

$$c_2 = \frac{3\sqrt{K_c(1 - K_c)K_0(1 - K_0)} + 3K_c(1 - K_0)}{2(K_c - K_0)}$$

$$c_3 = \max \begin{cases} \dfrac{2\sqrt{2}\hat{G}\sqrt{1 + 2K_c^2}}{K_c - 1} \\[3ex] \dfrac{-\sqrt{3}/\lambda^* + \sqrt{3}/\kappa^*}{2^{c_5\lambda^*} + \left(\dfrac{1}{50}\right)^{c_5\lambda^*} - 2} \end{cases}$$

$$c_4 = 1$$

$$c_5 = 1/K_c$$

$$c_6 = \frac{1}{2\left(\dfrac{-\sqrt{3}}{c_3\kappa^*} + 2^{c_5\lambda^*} - 1\right)}$$

$$\text{with} \quad K_c = \frac{1 - \sin\varphi_c}{1 + \sin\varphi_c}$$

$$K_0 = 1 - \sin\varphi_c$$

Table 5.2: Determination of constants c_1 - c_6. Note that c_5 has been determined by trial and error. Setting $c_5 = 1/K_c$ gives the best fit concerning an overall accordance.

Table 5.3: CSSM parameters used for calibration of barodesy

Material	φ_c	N	λ^*	κ^*	$\hat{G} = G_0/p_{\text{ini}}$	Parameters from[7]
London clay	22.6°	1.375	0.11	0.016	100*	Mašín [44]
Dresden clay	35°	0.622	0.038	0.008	60†	Medicus et al. [49]
Weald clay	24°	0.8	0.059	0.018	40	Mašín [47]
Dortmund clay	27.9°	0.749	0.057	0.008	30	Mašín [47]
Kaolin clay	27.5°	0.918	0.065	0.01	20	Mašín [47]
Fujinomori clay	34°	0.8867	0.0445	0.0108	30	Huang et al. [26]
San Francisco Bay Mud[8]	30.8°					Lade [40]

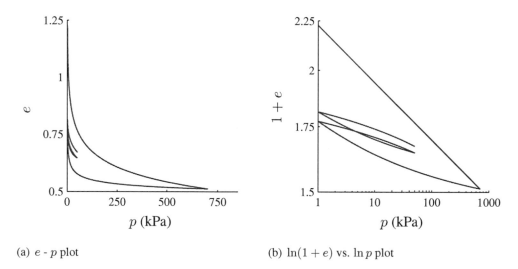

(a) e - p plot

(b) $\ln(1 + e)$ vs. $\ln p$ plot

Figure 5.6: Isotropic compression test on Weald Clay: Due to the underprediction of the unloading stiffness at low pressures, upheaval is simulated for certain stress amplitudes. In the plots, the loading process is $p = 1, 700, 1, 50, 1, 50$ kPa.

metric unloading is then overestimated. Note that for other stress loading/unloading ranges the resulting simulations are satisfactory, cf. the example in Figure 5.8 and the comparison of experimental data with barodesy in Section 5.3.

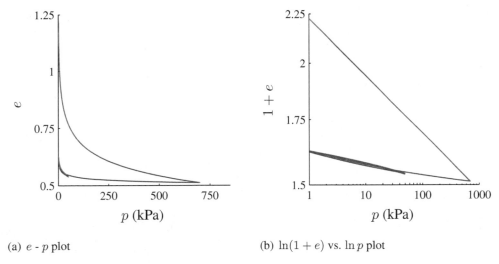

(a) e - p plot

(b) $\ln(1 + e)$ vs. $\ln p$ plot

Figure 5.7: Isotropic compression test on Weald Clay: Setting the parameter for the unloading stiffness to $\kappa^*/2$ reduces the upheaval effects at very low stresses. Note that for other stress amplitudes ratcheting is exhibited. In the plots, the loading process is chosen according to Figure 5.6

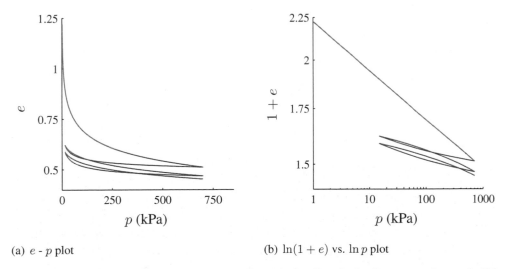

(a) e - p plot

(b) $\ln(1 + e)$ vs. $\ln p$ plot

Figure 5.8: Isotropic compression test on Weald Clay: In the plots, the loading process is $p = 1$, 700, 15, 700, 15, 700 kPa.

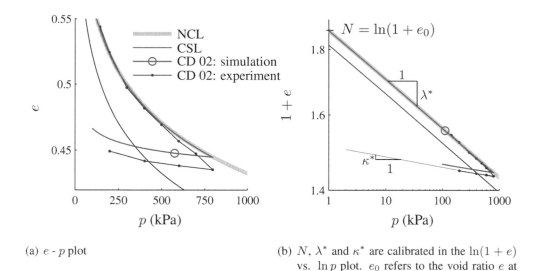

(a) e - p plot

(b) N, λ^* and κ^* are calibrated in the $\ln(1 + e)$ vs. $\ln p$ plot. e_0 refers to the void ratio e at $p = 1$ kPa.

Figure 5.9: Isotropic compression: Experimental results with Dresden clay CD 02 and simulations with barodesy.

5.3   Comparison with experimental data

In this Section, standard laboratory tests (isotropic, oedometric and triaxial (CD, CU) compression tests) are simulated with barodesy and are compared with experimental data for London clay, Dresden clay, Weald clay, Dortmund clay, Kaolin clay and Fujinomori clay, see Table 5.3.

5.3.1   Isotropic compression

As isotropic normal compression is explicitly included in the formulation and calibration of barodesy, normally consolidated isotropic loading test results are therefore in agreement with the simulated NCL, see Figures 5.9 - 5.11. As the unloading stiffness is described through the parameter κ^*, barodesy predicts satisfactorily results under isotropic unloading. In Section 2.2.1 (page 26) is described how the parameters N, λ^* and κ^* are calibrated. In Figure 5.9(b) the calibration is illustrated by using experimental data of Dresden clay.

In Figure 5.10 isotropic loading and unloading of Weald clay is shown. The normally and overconsolidated states marked in Figure 5.10(b) are initial states of triaxial tests shown in the next Section. Figure 5.11 shows an isotropic compression test of a slightly overconsolidated London clay sample. The asymptotic approach to the

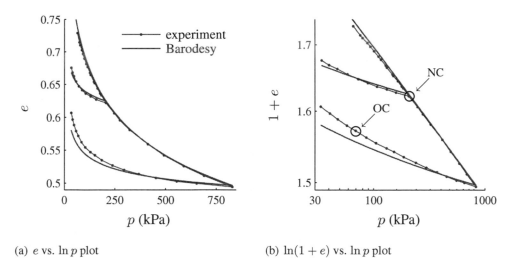

(a) e vs. $\ln p$ plot

(b) $\ln(1 + e)$ vs. $\ln p$ plot

Figure 5.10: Isotropic loading and unloading: Experimental results of Weald clay (according to Mašín [47], data by Henkel [21]) and numerical simulation with barodesy.

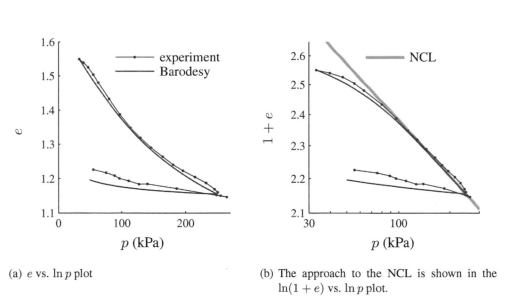

(a) e vs. $\ln p$ plot

(b) The approach to the NCL is shown in the $\ln(1 + e)$ vs. $\ln p$ plot.

Figure 5.11: Isotropic loading and unloading: Experimental results of London clay (Mašín [44]) and numerical simulation with barodesy.

NCL is visible. Thus, barodesy provides good predictions of isotropic compression behaviour of the clays analysed in this Section.

5.3.2   Triaxial compression

Critical strength is predicted with barodesy.[9] In Figure 5.12 limit points of normally consolidated samples obtained by true triaxial tests are shown. The data refer to San Francisco Bay Mud from Lade [40] and are compared with predictions by barodesy. The initial conditions for the triaxial tests in Figures 5.13 and 5.14 of the normally and overconsolidated samples are marked in Figure 5.10(b). In Figure 5.13 CU compression and extension tests of Weald clay and their simulations with barodesy are shown. The stress paths are satisfactorily predicted by barodesy (Figure 5.13a), the stress-strain behaviour slightly differs, see Figure 5.13(b). In Figure 5.14 drained triaxial tests are shown. The initial stiffnesses of the normally consolidated samples are slightly overpredicted, the predictions for the overconsolidated samples are satisfactory. In Figure 5.15 simulations of overconsolidated CU tests and experimental data of Dortmund clay (from Mašín [47] and Herle et al. [25]) are shown. The simultations with barodesy provide satisfactory results. In Figure 5.16 deviatoric triaxial compression of London clay with $p = 110$ kPa starting from a K_0-consolidated state with $e_{ini}=1.115$ (Mašín [44]) is simulated with barodesy. Peak strength is slightly underpredicted. Data of volumetric deformation are missing. Fujinomori clay drained triaxial data and their simulations are shown in Figures 5.17 - 5.19. The simulation overpredict the stiffness (apart from the initial stiffness). In Figure 5.20 deviatoric triaxial compression tests of Kaolin clay are shown. The overconsolidation ratios are between 1 and 50. The q vs. ε_q curves are predicted satisfactorily, cf. Figures 5.20(a), (b). Only the stiffness of the normally consolidated sample is overpredicted. In Figures 5.20(c), (d) the volumetric behaviour is shown. For the overconsolidated samples dilatancy is overpredicted by barodesy. The state boundary surface in Figure 5.20(e) is obtained by normalized representation of the triaxial tests. The experimental data of overconsolidated samples are reproduced accurately, for the normally consolidated test, q/p_e is overpredicted. As the volumetric behaviour of the normally consolidated sample is properly predicted, the variation in the SBS is therefore due to an overprediction of the stiffness, cf. Figures 5.20(b).

5.3.3   Oedometric compression

Oedometric compression of London clay is shown in Figure 5.21. The normal compression behaviour gives reasonable results, in the e vs. p plot as well as in the p vs. q plot.[10] For unloading, the radial stress is overpredicted, which results in an overestimation of p in the p-q as well as in the e-p plot.

---

[9]In Section 4.4.2 (page 53) is described how the critical strength is included in barodesy. The parameter φ_c can be determined relatively accurately.

[10]Compare also the predicted K_0-values in Figure 4.5(b).

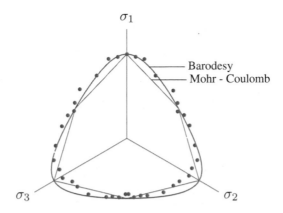

Figure 5.12: Critical stress points of normally consolidated San Francisco Bay Mud from Lade [40] are compared with critical state predictions by barodesy and Mohr-Coulomb. The calculations refer to $\mathrm{tr}\,\mathbf{T} = -500$ kPa and $\varphi_c = 30.6°$. The samples were isotropically consolidated and compressed in conventional and true triaxial tests. The critical stress points are arranged slightly anisotropic, due to anisotropic orientation of the particles, according to Lade [40]. An isotropic material would leave a rotation in the deviatoric plane by $120°$ undiscovered.

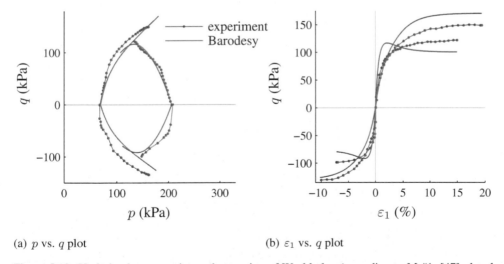

(a) p vs. q plot

(b) ε_1 vs. q plot

Figure 5.13: Undrained compression and extension of Weald clay (according to Mašín [47], data by Parry [56]) and numerical simulation with barodesy. The initial states of the normally and overconsolidated samples are marked in Figure 5.10(b).

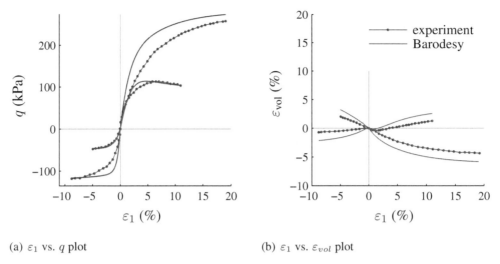

(a) ε_1 vs. q plot

(b) ε_1 vs. ε_{vol} plot

Figure 5.14: Drained compression and extension of Weald clay (according to Mašín [47], data by Parry [56]) and numerical simulation with barodesy.

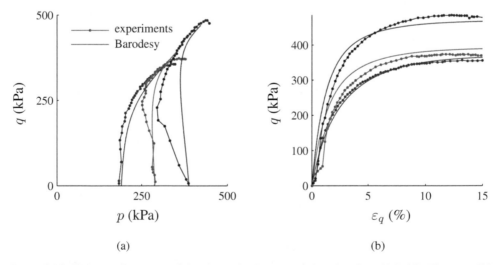

(a)

(b)

Figure 5.15: CU tests of overconsolidated samples Dortmund clay, data from [25, 47]; The consolidation pressures for the samples are $p_{ini} = 190$ kPa, $p_{ini} = 290$ kPa and $p_{ini} = 390$ kPa with the void ratios $e_{ini} = 0.46$, $e_{ini} = 0.455$ and $e_{ini} = 0.44$.

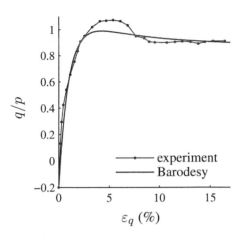

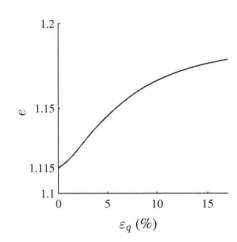

(a) q/p vs. ε_q plot

(b) e vs. ε_q plot; experimental results of London clay are missing.

Figure 5.16: Deviatoric triaxial compression of London clay with $p = 110$ kPa $=$ const starting from K_0-consolidated state with $e_{\mathrm{ini}} = 1.115$ (data from Mašín [44]) and numerical simulation with barodesy.

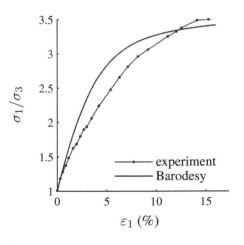

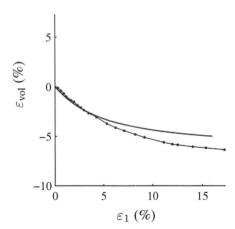

(a) σ_1/σ_3 vs. ε_1 plot

(b) $\varepsilon_{\mathrm{vol}}$ vs. ε_1 plot

Figure 5.17: Triaxial compression of Fujinomori clay, according to Huang et al. [26] with data from Nakai et al. [52] with $\sigma_3 = 196$ kPa $e_{\mathrm{ini}} = 0.915$ and numerical simulation with barodesy.

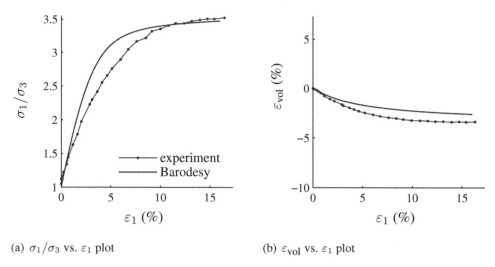

(a) σ_1/σ_3 vs. ε_1 plot

(b) ε_{vol} vs. ε_1 plot

Figure 5.18: Triaxial compression of Fujinomori clay, according to Huang et al. [26] with data from Nakai et al. [52] with $p = 196$ kPa = const, $e_{\text{ini}} = 0.915$ and numerical simulation with barodesy.

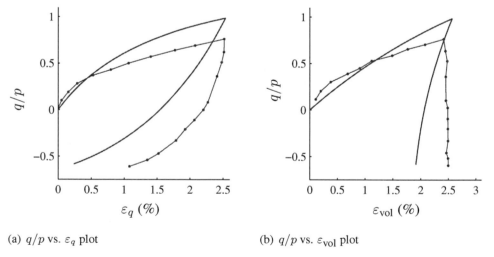

(a) q/p vs. ε_q plot

(b) q/p vs. ε_{vol} plot

Figure 5.19: Triaxial compression of Fujinomori clay, according to Huang et al. [26] with data from Nakai et al. [52] with $\sigma_3 = 196$ kPa = const, $e_{\text{ini}} = 0.915$ and numerical simulation with barodesy.

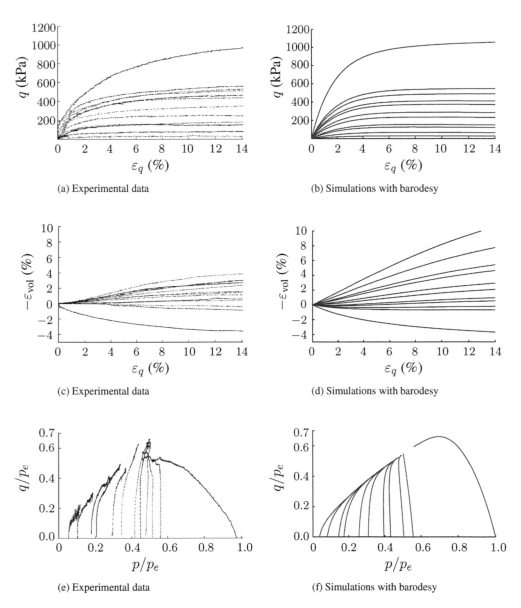

(a) Experimental data

(b) Simulations with barodesy

(c) Experimental data

(d) Simulations with barodesy

(e) Experimental data

(f) Simulations with barodesy

Figure 5.20: Kaolin clay: CD tests with $p = $ const: q - ε_q plots, ε_{vol} - ε_q plots and state boundary surface. The experimental data in Figures (a), (c), (e) are taken from Mašín [47] with data from Hattab & Hicher [20].

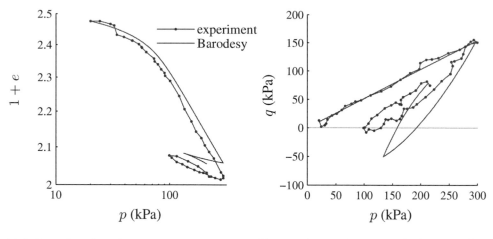

(a) $\ln(1+e)$ vs. $\ln p$ plot

(b) p - q plot: σ_2 (p respectively) is overpredicted at unloading.

Figure 5.21: Oedometric loading (up to $\sigma_1 = 400$ kPa), unloading (up to $\sigma_1 = 100$ kPa) and reloading ($\sigma_1 = 266$ kPa): Experimental results (*PhM14*) of London Clay (data from Mašín [43]) and numerical simulation with barodesy

In this Section experimental results of different clays are compared with simulations of barodesy. Testing methods are limited to certain standard tests as oedometric/isotropic compression tests, triaxial tests and a few more. It is important to show that basic trends are correctly predicted. Therefore basic features of various clays[11] are summarized, in order to gain an overview of how certain parameters influence the model's behaviour and in order to check the model's response for different soils. In Figure 5.22 basic features are shown exemplary for London clay. For simulations of Dresden clay, Weald clay, Dortmund clay, Kaolin clay and Fujinomori clay, we refer to Figures E.1- E.5 in Appendix E. For the simulations in Figures 5.22 and E.1- E.5 the following conditions are chosen: The initial stress state for Figures (a) and (d) is $\mathbf{T}_{\mathrm{ini}} = -50 \cdot \mathbf{1}$ kPa. The initial void ratio in Figures (a), (c) and (d) is $e_{\mathrm{ini}} = \exp\left(N - \lambda^* \ln(2p/\sigma^*)\right) - 1$. Figures (a) show the state boundary surfaces for different clays. Their qualitative shapes seem reasonable. In Figures (b) Chu & Lo's stress-dilatancy relation is compared with predictions by barodesy. For soils with relatively high critical friction angles[12], strength is slightly underpredicted for dilatant states (i.e. $\tan \beta > 0$), whereas for low friction angles[13], strength is slightly overpredicted. The K_0 predictions by Chu & Lo, Jáky and barodesy lie close together for all investigated soils. In Figures (c) the asymptotic approach to PσPs is

---

[11]The material parameters of the analysed clays are in Table 5.3.

[12]For Dresden clay φ_c equals $35°$, for Fujinomori clay φ_c equals $34°$.

[13]For London clay φ_c equals $22.6°$, for Weald clay φ_c equals $24°$.

shown. Several response envelopes are shown in Figures (d). The initial states are arbitrary permissible stress states. In Figures (e) and (f) soil is isotropically consolidated up to $p = 500$ kPa, unloaded to $p = 10$ kPa and reloaded to 1000 kPa. In Figures 5.22(e) and 5.22(f) the hysteretic loop is visible. Note that wheater the hysteretic loop appears, depends on the stress amplitude, as well as on the material parameters (see Table 5.3).

Another possibility to estimate the model's performance, is to compare it with other models. The hypoplastic model by Mašín and the Modified Cam Clay model are chosen as references (see Chapter 3 on page 41). In Figure 5.23 simulations of isotropic compression tests are compared with experimental data. In Figure 5.24, simulations of triaxial tests with constant mean pressure starting from K_0 - consolidated states are shown. Peak strength is slightly underpredicted with barodesy and overpredicted with the MCC-model. The model by Mašín gives the best fit of peak strength.

Jáky and Chu & Lo (equation 2.7) make predictions of stress states under oedometric normal compression. In Figure 5.25 their predictions are compared with the MCC-model, the Mašín-model and barodesy. Barodesy and the current model by Mašín [47] meet K_0 according to Jáky and Chu & Lo best.[16] The MCC-model strongly overestimates the K_0-values. The model by Mašín [44] slightly overestimates the K_0 values.

In Figures 5.26 and 5.27 drained triaxial tests are simulated with barodesy and the hypoplastic version of Mašín. The triaxial tests are shown as p vs. e and p vs. q plots as well as plots in the normalized stress plane (i.e. p/p_e vs. q/p_e).

---

[15]The peak value of the triaxial test follows from the initial yield locus with $p_0 = \exp\left(\dfrac{N - \ln(1 + e_{\text{ini}}) - \kappa^* \ln p}{\lambda^* - \kappa^*}\right)$, $e_{\text{ini}} = 1.115$ and $p = 110$ kPa.

[16]Note that Jáky's equation is included in barodesy and in the current model by Mašín [47] and therefore practically agrees with $K_0 = 1 - \sin\varphi_c$.

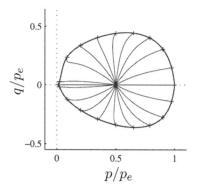

(a) State boundary surface

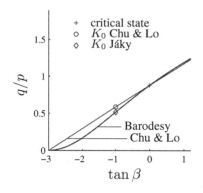

(b) Chu & Lo's predictions are compared
      with barodesy

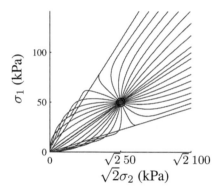

(c) Starting from $\mathbf{T} \neq \mathbf{0}$ and applying
      PεPs

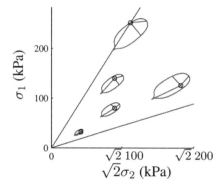

(d) Response envelopes

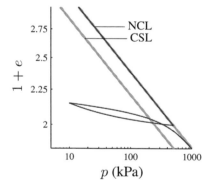

(e) Loading, unloading and reloading un-
      der isotropic compression in the $\ln p$ -
      $\ln(1 + e)$ plot

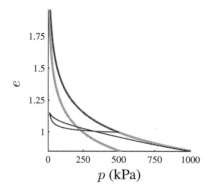

(f) Loading, unloading and reloading un-
      der isotropic compression in the p - e
      plot

Figure 5.22: London clay behaviour, simulated with barodesy

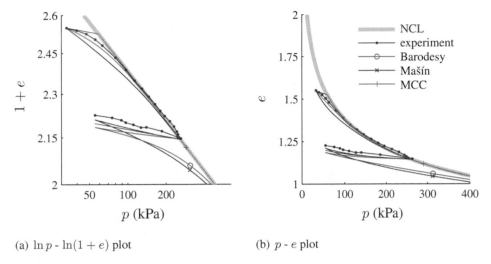

(a) $\ln p$ - $\ln(1 + e)$ plot

(b) p - e plot

Figure 5.23: Isotropic compression test of London clay: Numerical simulations obtained with barodesy, hypoplastic model of Mašín [44] and the Modified Cam Clay model; London Clay, data from [44]. Reloading is simulated, even though experimental data are missing.

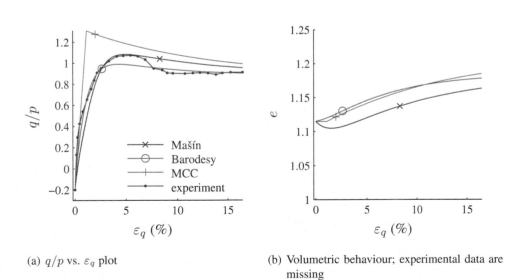

(a) q/p vs. ε_q plot

(b) Volumetric behaviour; experimental data are missing

Figure 5.24: Deviatoric triaxial compression tests ($p = 110$ kPa $=$ const, $e_{ini} = 1.115$): Numerical simulations obtained with barodesy, hypoplastic model of [44] and the Modified Cam Clay model[15]; London Clay, data from [44]

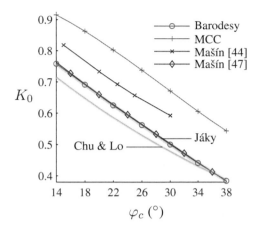

Figure 5.25: Oedometric normal compression: Chu & Lo's prediction and Jáky's equation $K_0 = 1 - \sin \varphi_c$ are compared with predictions by barodesy, with the Modified Cam Clay model (MCC), as well as the hypoplastic version by Mašín [44] and *Clay hypoplasticity with explicitly defined asymptotic states* by Mašín [47]; Barodesy satisfies Jáky's equation and Mašín [47] approximately satisfies Jáky's equation. MCC and Mašín [44]: data from Mašín & Herle [48].

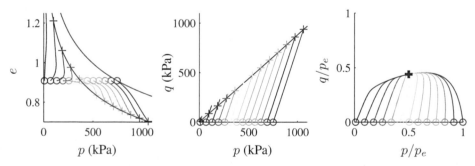

Figure 5.26: Barodesy: Simulated paths of drained triaxial tests for London Clay. The start points are denoted by a circle, the end points are denoted by a cross

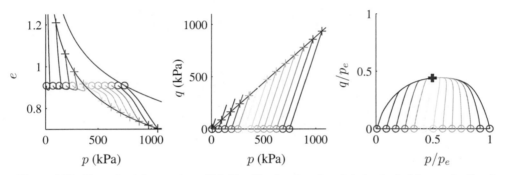

Figure 5.27: Hypoplasticity version of Mašín: Simulated paths of drained triaxial tests for London Clay. The start points are denoted by a circle (o), the end points are denoted by a cross (+)

Chapter 6

Concluding remarks

Summary

The asymptotic behaviour of soil deserves particular attention, as Gudehus pointed to. It provides the basis of barodesy, which includes important characteristics of soil behaviour, such as critical states, barotropy, pyknotropy and a stress-dilatancy relation. Barodesy can be written as a single equation of the form $\dot{\mathbf{T}} = \mathbf{h}(\mathbf{T}, \mathbf{D}, e)$, i.e. the stress rate is expressed as a function of the stress, stretching and the void ratio.

The here presented constitutive relation has been compared with

  i the hypoplastic constitutive model for clays by Mašín,

  ii the Modified Cam Clay model,

  iii clay hypoplasticity with explicitly defined asymptotic states by Mašín,

  iv Rowe's relation and

  v Chu & Lo's relation.

A basic ingredient of barodesy is the relation $\mathbf{R}(\mathbf{D})$. The $\mathbf{R}$-function (equation 4.2) connects stretchings with the corresponding proportional stress paths, and thus includes a stress-dilatancy relation. In the course of this work the mathematical formulation of $\mathbf{R}(\mathbf{D})$ was simplified and further improved. I have compared the $\mathbf{R}$-function with the stress-dilatancy relations of (ii), (iii), (iv) and (v). The results of barodesy and (iii) almost agree and perform best compared with the other relations, which apply only for a small range of stretching directions. In order to calibrate the $\mathbf{R}$-function, only the critical friction angle is required. The introduced relation $\mathbf{R}$ is applicable for sand *and* clay.

I have focussed on a modification of barodesy in order to link Critical State Soil Mechanics with barodesy and to make it thus applicable for clay. I also have developed a calibration method: Five well-known material parameters of Soil Mechanics, which can be determined from a consolidated undrained compression test, are sufficient. The model provides realistic results, as compared with experimental results of various clay types.

Table 5.1 (page 70) summarizes the mathematical formulation of the model, in Table 5.2 the determination of the material constants is presented. Table 5.3 gives an overview of material parameters for various clay types. Compared with other constitutive relations, barodesy is simple and easy to understand and this offers potential to further development.

Conclusions, limitations and outlook

Conclusions

1. The exponential function for $\mathbf{R}$ is suitable for general, i.e. non axisymmetric, paths and can be adapted to existing stress-dilatancy relations.

2. Barodesy is applicable for clay, too.

Limitations

1. As in hypoplasticity, barodesy uses only $\mathbf{T}$ and e as memory parameters. This covers many phenomena, is however insufficient to capture all memory effects. Consequently, ratcheting and unrealistic small-strain behaviour are obtained.

2. At very low stress levels, the unloading stiffness is underestimated, cf. the isotropic compression test in Figure 5.6 on page 74.

3. For certain loading/unloading ranges, upheavals are obtained, cf. the isotropic compression test in Figure 5.6 on page 74.

Modification of f and g could improve some simulations. The easiest way to increase the unloading stiffness is to decrease the parameter κ^*, e.g. by the factor $1/2$, cf. Figure 5.7 on page 75. Upheavals are then reduced, but the unloading stiffness close to the NCL is overestimated. The overall performance of the model would not be much affected. As shown in Section 5.3, for other stress loading/unloading ranges the resulting simulations for isotropic and oedometric tests are satisfactory, cf. the example in Figure 5.8 on page 75.

Outlook

1. A goal is to develop a simple and robust calibration method for sand, similar to the one for clay.

2. A user-defined material subroutine of ABAQUS is available for an earlier version of barodesy (Medicus et al. [49]). A FEM implementation of the actual version will soon enable practical applications.

3. Modelling rate-dependence and relaxation, as well as cylic loading[1] is left for future.

---

[1]Kolymbas [39] proposes to capture small strain stiffness and cyclic loading by considering rate dependence.

Bibliography

[1] J. H. Atkinson. *An Introduction to the Mechanics of Soils and Foundations.* McGraw Hill, 1993.

[2] J. H. Atkinson and P. L. Bransby. *The Mechanics of Soils: An Introduction to Critical State Soil Mechanics.* McGraw-Hill Book Co., 1978.

[3] E. Bauer. Constitutive modelling of critical states in hypoplasticity. In *Proceedings of the Fifth International Symposium on Numerical Models in Geomechanics, Davos, Switzerland*, pages 15–20. Balkema, 1995.

[4] E. Bauer. Calibration of a comprehensive hypoplastic model for granular materials. *Soils and Foundations*, 36(1):13–26, 1996.

[5] M. D. Bolton. The strength and dilatancy of sands. *Géotechnique*, 36:65–78, 1986.

[6] R. Butterfield. A natural compression law for soils (an advance on e-log p'). *Géotechnique*, 29(4):469 – 480, 1979.

[7] L. Callisto and G. Calabresi. Mechanical behaviour of a natural soft clay. *Géotechnique*, 48:495–513(18), 1998.

[8] J. Chu and S.-C. R. Lo. Asymptotic behaviour of a granular soil in strain path testing. *Géotechnique*, 44(1):65–82, 1994.

[9] J. Desrues, B. Zweschper, and P. Vermeer. Database for Tests on Hostun RF Sand. Technical report, Institute for Geotechnical Engineering, University of Stuttgart, 2000.

[10] W. Fellin and A. Ostermann. The critical state behaviour of barodesy compared with the Matsuoka–Nakai failure criterion. *International Journal for Numerical and Analytical Methods in Geomechanics*, 37(3):299–308, 2013. doi: 10.1002/nag.1111.

[11] A. Gajo and D. Muir Wood. Severn-Trent Sand: a kinematic-hardening constitutive model: the q-p formulation. *Géotechnique*, 49(5):696–614, 1999.

[12] A. Gajo and D. Muir Wood. A kinematic hardening constitutive model for sands: the multiaxial formulation. *International Journal for Numerical and Analytical Methods in Geomechanics*, 23(9):925–965, 1999. ISSN 1096-9853.

95

[13] A. Gasparre. *Advanced laboratory characterisation of London clay.* PhD thesis, Imperial College of Science and Technology, London, 2005.

[14] M. Goldscheider. Grenzbedingung und Fließregel von Sand. *Mechanics Research Communications*, 3:463–468, 1967.

[15] G. Gudehus. A comparison of some constitutive laws for soils under radially symmetric loading and unloading. In Wittke, editor, *Proc., 3rd Conf. Numerical Methods in Geomechanics*, pages 1309–1323, 1979.

[16] G. Gudehus. A comprehensive constitutive equation for granular materials. *Soils and Foundations*, 36(1):1–12, 1996.

[17] G. Gudehus. *Physical Soil Mechanics.* Springer, Berlin, 2011.

[18] G. Gudehus and D. Mašín. Graphical representation of constitutive equations. *Géotechnique*, 59(2):147–151, 2009.

[19] G. Gudehus, M. Goldscheider, and H. Winter. Mechanical properties of sand and clay and numerical integration methods: some sources of errors and bounds of accuracy. In G. Gudehus, editor, *Finite Elements in Geomechanics*, pages 121–150. John Wiley & Sons, 1977.

[20] M. Hattab and P. Hicher. Dilating behaviour of overconsolidated clay. *Soils and Foundations*, 44(4):27–40, 2004.

[21] D. Henkel. The effect of overconsolidation on the behaviour of clays during shear. *Géotechnique*, 6:139–150, 1956.

[22] I. Herle. *Hypoplastizität und Granulometrie einfacher Korngerüste.* PhD thesis, Veröffentlichungen des Institutes für Bodenmechanik und Felsmechanik der Universität Karlsruhe (No. 142), 1997.

[23] I. Herle and G. Gudehus. Determination of parameters of a hypoplastic constitutive model from properties of grain assemblies. *Mechanics of cohesive-frictional materials*, 4:461–486, 1999.

[24] I. Herle and D. Kolymbas. Hypoplasticity for soils with low friction angles. *Computers and Geotechnics*, 31(5):365–73, 2004.

[25] I. Herle, D. Mašín, V. Kostkanová, C. Karcher, and D. Dahmen. Experimental investigation and theoretical modelling of soft soils from mining deposits. *Proceedings of the 5th international symposium on deformation characteristics of geomaterials*, 2:858–864, 2011.

[26] W.-X. Huang, W. Wu, D.-A. Sun, and S. Sloan. A simple hypoplastic model for normally consolidated clay. *Acta Geotechnica*, 1:15–27, 2006. ISSN 1861-1125. doi: 10.1007/s11440-005-0003-3.

[27] N. Janbu. Soil compressibility as determined by oedometer and triaxial tests. In *Proceedings of the European Conference on Soil Mechanics and Foundation Engineering*, 1963.

[28] S. Jänke. *Untersuchung der Zusammendrückbarkeit und Scherfestigkeit von Sanden und Kiesen sowie der sie bestimmenden Einflüsse*. Mitteilungsblatt der Bundesanstalt für Wasserbau. Bundesanstalt für Wasserbau, 1969.

[29] N. Khalili, M. A. Habte, and S. Valliappan. A bounding surface plasticity model for cyclic loading of granular soils. *International Journal for Numerical Methods in Engineering*, 63(14):1939–1960, 2005. ISSN 1097-0207. doi: 10.1002/nme.1351.

[30] D. Kolymbas. A rate-dependent constitutive equation for soils. *Mech. Res. Commun*, 4:367–372, 1977.

[31] D. Kolymbas. A generalized hypoelastic constitutive law. *Proceedings of the 11th international conference on soil mechanics and foundation engineering, San Francisco, Balkema*, 5:2626, 1985.

[32] D. Kolymbas. *Eine konstitutive Theorie für Böden und andere körnige Stoffe*. PhD thesis, Veröffentlichungen des Institutes für Bodenmechanik und Felsmechanik der Universität Karlsruhe (Heft 109), 1988.

[33] D. Kolymbas. An outline of hypoplasticity. *Archive of Applied Mechanics*, 61(3):143–151, 1991. doi: 10.1007/BF00788048.

[34] D. Kolymbas. Hypoplasticity as a constitutive framework for granular materials. In Siriwardane and Zaman, editors, *Computer Methods and Advances in Geomechanics*, pages 197–208. Balkema, 1994.

[35] D. Kolymbas. *Introduction to Hypoplasticity*. Number 1 in Advances in Geotechnical Engineering and Tunnelling. Balkema, Rotterdam, 2000.

[36] D. Kolymbas. Sand as an archetypical natural solid. In D. Kolymbas and G. Viggiani, editors, *Mechanics of Natural Solids*, pages 1–26. Springer: Berlin, 2009.

[37] D. Kolymbas. Barodesy: a new constitutive frame for soils. *Géotechnique Letters*, 2:17–23, 2012. doi: 10.1680/geolett.12.00004.

[38] D. Kolymbas. Barodesy: a new hypoplastic approach. *International Journal for Numerical and Analytical Methods in Geomechanics*, 36(9):1220–1240, 2012. ISSN 1096-9853. doi: 10.1002/nag.1051.

[39] D. Kolymbas. Barodesy as a novel hypoplastic constitutive theory based on the asymptotic behaviour of sand. *Geotechnik*, 35(3):187–197, 2012. ISSN 2190-6653. doi: 10.1002/gete.201200002.

[40] P. V. Lade. Effects of consolidation stress state on normally consolidated clay. In H. Rathmayer, editor, *Proceedings of NGM-2000: XIII Nordiska Geoteknikermötet: Helsinki, Finland*. Building Information Ltd., 2000.

[41] S. Leroueil, M. Roy, P. L. Rochelle, F. Brucy, and F. A. Tavenas. Behavior of destructured natural clays. *Journal of the Geotechnical Engineering Division*, 106(6):759–778, 1979.

[42] T. Luo, Y.-P. Yao, A.-N. Zhou, and X.-G. Tian. A three-dimensional criterion for asymptotic states. *Computers and Geotechnics*, 41(0):90 – 94, 2012. ISSN 0266-352X. doi: 10.1016/j.compgeo.2011.12.002.

[43] D. Mašín. *Laboratory and numerical modelling of natural clay*. MPhil thesis, City University, London, 2004.

[44] D. Mašín. A hypoplastic constitutive model for clays. *International Journal for Numerical and Analytical Methods in Geomechanics*, 29(4):311–336, 2005.

[45] D. Mašín. Hypoplastic Cam-clay model. *Géotechnique*, 62(6):549–553, 2012.

[46] D. Mašín. Asymptotic behaviour of granular materials. *Granular Matter*, 14: 759–774, 2012. ISSN 1434-5021.

[47] D. Mašín. Clay hypoplasticity with explicitly defined asymptotic states. *Acta Geotechnica*, 8(5):481–496, 2013. ISSN 1861-1125. doi: 10.1007/s11440-012-0199-y.

[48] D. Mašín and I. Herle. State boundary surface of a hypoplastic model for clays. *Computers and Geotechnics*, 32(6):400 – 410, 2005. ISSN 0266-352X. doi: 10.1016/j.compgeo.2005.09.001.

[49] G. Medicus, W. Fellin, and D. Kolymbas. Barodesy for clay. *Géotechnique Letters*, 2:173–180, 2012. doi: 10.1680/geolett.12.00037.

[50] D. Muir Wood. *Soil Behaviour and Critical State Soil Mechanics*. Cambridge University Press, 1990.

[51] D. Muir Wood. *Soil Mechanics: A One-Dimensional Introduction*. Cambridge University Press, 2009.

[52] T. Nakai, H. Matsuoka, N. Okuno, and K. Tsuzuki. True triaxial tests on normally consolidated clay and analysis of the observed shear behaviour using elastoplastic constitutive models. *Soils and Foundations*, 26(4):67–78, 1986.

[53] A. Niemunis. *Extended hypoplastic models for soils*. Number 1 in Schriftenreihe des Institutes für Grundbau und Bodenmechanik der Ruhr-Universität Bochum. Habilitation Thesis, 2003.

[54] A. Niemunis and I. Herle. Hypoplastic model for cohesionless soils with elastic strain range. *Mechanics of Cohesive-Frictional Materials*, 2(4):279–299, 1997.

[55] J. Ohde. Zur Theorie der Druckverteilung im Baugrund. *Der Bauingenieur*, 20:451–459, 1939.

[56] R. Parry. Triaxial compression and extension tests on remoulded saturated clay. *Géotechnique*, 10(4):166–180, 1960.

[57] S. Rampello and L. Callisto. A study on the subsoil of the Tower of Pisa based on results from standard and high-quality samples. *Canadian Geotechnical Journal*, 35(6):1074–1092, 1998.

[58] K. Roscoe and J. Burland. On the generalised stress-strain behaviour of wet clay. In J. Heyman and F. Leckie, editors, *Engineering Plasticity*, pages 535–609. Cambridge University Press: Cambridge, 1968.

[59] P. Rowe. The stress-dilatancy relation for static equilibrium of an assembly of particles in contact. In *Proceedings of the Royal Society of London. Series A, Mathematical and Physical Sciences*, volume 296, pages 500–527, 1962.

[60] M. Taiebat and Y. F. Dafalias. Sanisand: Simple anisotropic sand plasticity model. *International Journal for Numerical and Analytical Methods in Geomechanics*, 32(8):915–948, 2008. ISSN 1096-9853. doi: 10.1002/nag.651.

[61] D. Taylor. *Fundamentals of soils mechanics*. John Wiley & Sons, New York, 1948.

[62] M. Topolnicki. *Observed stress-strain behaviour of remoulded saturated clay and examination of two constitutive models*. PhD thesis, Veröffentlichungen des Institutes für Bodenmechanik und Felsmechanik der Universität Karlsruhe (No. 107), 1987.

[63] M. Topolnicki, G. Gudehus, and B. Mazurkiewicz. Observed stress-strain behaviour of remoulded saturated clays under plane strain conditions. *Géotechnique*, 40(2):155–187, 1990. doi: 10.1680/geot.1990.40.2.155.

[64] P.-A. von Wolffersdorff. A hypoplastic relation for granular materials with a predefined limit state surface. *Mechanics of Cohesive-Frictional Materials, 1*, 1:251–271, 1996.

[65] W. Wu and E. Bauer. A hypoplastic model for barotropy and pyknotropy of granular soils. In D. Kolymbas, editor, *Modern Approaches to Plasticity*. Elsevier, 1993.

[66] W. Wu and E. Bauer. A simple hypoplastic constitutive model for sand. *International Journal for Numerical and Analytical Methods in Geomechanics*, 18: 833–862, 1994.

[67] W. Wu and D. Kolymbas. Numerical testing of the stability criterion for hypoplastic constitutive equations. *Mechanics of Materials*, 9:245:253, 1990.

Appendix A

List of symbols

In this appendix symbols which are often used are summarized. The pages referred to, denote either the page where the symbol was introduced or the page where the symbol is described best.

Symbol	Meaning
c_i	material constants
e	void ratio, page 4
e_c	pressure dependent critical void ratio, page 9
e_i, e_d	pressure dependent maximum and minimum void ratio, pages 36, 37
e_{i0}, e_{c0}, e_{d0}	maximum, critical and minimum void ratio at zero stress level, page 39
e_{ini}	initial void ratio
e_{max}, e_{min}	maximum and minimum void ratio, according to index tests
f	scalar quantity, part of barodesy, page 48
f	Only in the context of elasto-plasticity, f is the yield function, page 41.
g	scalar quantity, part of barodesy, page 48
g	Only in the context of elasto-plasticity, g is the plastic potential, page 42.
f_d	factor of pyknotropy, page 2
f_s	factor of barotropy, page 2
m	dilation measure, ratio of volumetric to deviatoric strain increment $\dot{\varepsilon}_{\text{vol}}/\dot{\varepsilon}_q$, page 5
p	mean effective pressure, $p = 1/3\,(\sigma_1 + \sigma_2 + \sigma_3)$, page 4
p_{ini}	initial mean effective pressure
p_0	preloading mean effective pressure
p_e	Hvorslev's equivalent consolidation pressure, page 32
q	deviatoric stress, for axisymmetric conditions $q = \sigma_1 - \sigma_2$, page 4
E_s	incremental stiffness under oedometric compression $E_s = \dot{\sigma}_1/\dot{\varepsilon}_1$, page 10
G	shear modulus, page 11

H hardening parameter, page 42
I_d density index, page 24
I_r relative density index, page 24
K ratio of radial to axial stress, page 50
K_0 ratio of radial to axial stress under normal oedometric compression, page 14
K_c ratio of radial to axial stress at critical state, for compression $K_c = (1 - \sin \varphi_c)/(1 + \sin \varphi_c)$, page 14
M slope of the CSL in p-q-plot, for compression $M = 6 \sin \varphi_c/(3 - \sin \varphi_c)$, page 29
N ordinate intercept of the NCL, page 26

$\mathbf{D}$ stretching, page 4
$\mathbf{D}_e$ elastic stiffness matrix, page 41
$\mathbf{D}_{ep}$ elasto-plastic stiffness matrix, page 41
$\mathcal{L}$ fourth-order constitutive tensor, page 45
$\mathbf{N}$ second-order constitutive tensor, page 45
$\mathbf{R}$ $\mathbf{R}$-function, defining the direction of proportional paths (PσPs), page 47
$\mathbf{T}$ effective Cauchy stress, page 1
$\dot{\mathbf{T}}$ effective Cauchy stress rate, page 4
$\overset{\circ}{\mathbf{T}}$ objective rate of the effective Cauchy stress, $\overset{\circ}{\mathbf{T}} = \dot{\mathbf{T}}$ for rectilinear extensions, page 1
$\mathbf{X}^0$ normalized tensor $\mathbf{X}^0 = \mathbf{X}/|\mathbf{X}|$
$|\mathbf{X}|$ Euclidean norm of a tensor $|\mathbf{X}| = \sqrt{\mathrm{tr}\,\mathbf{X}^2}$

α scalar quantity, part of the $\mathbf{R}$-function, page 47
β scalar quantity, part of barodesy functions f and g, page 49
$\tan \beta$ dilation measure, ratio of volumetric to axial strain increment $-\dot{\varepsilon}_{\mathrm{vol}}/\dot{\varepsilon}_1$, page 4
ε_q deviatoric strain, page 4
ε_i principal strain, for axisymmetric conditions ε_1 is used for the axial strain and $\varepsilon_2 = \varepsilon_3$ is used for the radial strain
$\varepsilon_{\mathrm{vol}}$ volumetric strain, page 4
δ dilaton measure $\delta = \mathrm{tr}\,\mathbf{D}^0 = -\dot{\varepsilon}_{\mathrm{vol}}/|\dot{\varepsilon}|$, page 4
$\dot{\varepsilon}$ Euclidean norm of $\mathbf{D}$, $\dot{\varepsilon} = \sqrt{\mathrm{tr}\,\mathbf{D}^2}$, page 4
κ, κ^* slope of the unloading line under isotropic compression in the $\ln p$ - $(1 + e)$ plot, $\ln p$ - $\ln(1 + e)$ plot, respectively, page 26
λ, λ^* slope of the NCL in the $\ln p$ - $(1 + e)$ plot, $\ln p$ - $\ln(1 + e)$ plot, respectively, page 26
Λ scalar quantity, part of barodesy, page 66

σ	Euclidean norm of $\mathbf{T}$, $\sigma = \sqrt{\operatorname{tr} \mathbf{T}^2}$, page 4
σ^*	reference stress $\sigma^* = 1$ kPa, page 48
σ_i	principal effective stress, for axisymmetric conditions σ_1 is used for the axial effective stress and $\sigma_2 = \sigma_3$ is used for the radial effective stress
φ_c	critical friction angle, page 9
φ_m	mobilised friction angle, page 20
φ_p	peak friction angle, page 24
$\psi_{\dot{\varepsilon}}$	angle, defining the position of a PεP in the Rendulic plane ε_1 - $\sqrt{2}\varepsilon_2$, page 18
ψ_σ	angle, defining the position of a PσP in the Rendulic plane σ_1 - $\sqrt{2}\sigma_2$, page 18

Appendix B

Equations of the hypoplastic model by Mašín [44]

$$\dot{\mathbf{T}} \;=\; f_s \mathcal{L} \mathbf{D} + f_s f_d \mathbf{N} |\mathbf{D}| \tag{B.1}$$

$$\text{with: } \mathbf{N} \;=\; \mathcal{L} : \left(-Y \frac{\mathbf{m}}{|\mathbf{m}|} \right)$$

$$\mathcal{L} \;=\; 3 \left(c_1 \mathcal{I} + c_2 a^2 \hat{\mathbf{T}} \otimes \hat{\mathbf{T}} \right)$$

Stress invariants:
$$I_1 \;=\; \operatorname{tr} \mathbf{T}, \; I_2 = \frac{1}{2}[\mathbf{T} : \mathbf{T} - I_1^2], \; I_3 = \det \mathbf{T}$$

$$\hat{\mathbf{T}} \;=\; \frac{\mathbf{T}}{\operatorname{tr} \mathbf{T}}, \quad \hat{\mathbf{T}}^* = \hat{\mathbf{T}} - \frac{1}{3}$$

$$\tan \psi \;=\; \sqrt{3 \operatorname{tr} \hat{\mathbf{T}}^{*2}}, \quad \cos 3\vartheta = -\sqrt{6} \frac{\operatorname{tr} \hat{\mathbf{T}}^{*3}}{\left[\operatorname{tr} \hat{\mathbf{T}}^{*2} \right]^{3/2}}$$

$$F \;=\; \sqrt{\frac{1}{8} \tan^2 \psi + \frac{2 - \tan^2 \psi}{2 + \sqrt{2} \tan \psi \cos 3\vartheta}} - \frac{1}{2\sqrt{2}} \tan \psi$$

Scalar quantities:
$$a \;=\; \frac{\sqrt{3}(3 - \sin \varphi_c)}{2\sqrt{2} \sin \varphi_c}$$

$$c_1 \;=\; \frac{2(3 + a^2 - 2^\alpha a\sqrt{3})}{9r}$$

$$c_2 \;=\; 1 + (1 - c_1)\frac{3}{a^2}$$

$$\alpha \;=\; \frac{1}{\ln 2} \ln \left[\frac{\lambda^* - \kappa^*}{\lambda^* + \kappa^*} \left(\frac{3 + a^2}{a\sqrt{3}} \right) \right]$$

Limit stress condition:
$$Y \;=\; \left(\frac{\sqrt{3}a}{3 + a^2} - 1 \right) \frac{(I_1 I_2 + 9 I_3)(1 - \sin^2 \varphi_c)}{8 I_3 \sin^2 \varphi_c} + \frac{\sqrt{3}a}{3 + a^2}$$

Hypoplastic flow rule: $\quad \mathbf{m} \quad = \quad -\dfrac{a}{F}\left[\hat{\mathbf{T}} + \hat{\mathbf{T}}^* - \dfrac{\hat{\mathbf{T}}}{3}\left(\dfrac{6\hat{\mathbf{T}}:\hat{\mathbf{T}} - 1}{(F/a)^2 + \hat{\mathbf{T}}:\hat{\mathbf{T}}}\right)\right]$

Barotropy factor: $\quad f_s \quad = \quad -\dfrac{\mathrm{tr}\,\mathbf{T}}{\lambda^*}(3 + a^2 - 2^\alpha a\sqrt{3})^{-1}$

Pyknotropy factor: $\quad f_d \quad = \quad \left(-\dfrac{2\mathrm{tr}\,\mathbf{T}}{3p_r}\exp\left(\dfrac{\ln(1+e) - N}{\lambda^*}\right)\right)^\alpha$

Appendix C

Barodesy in the case of axial symmetry

Most element test exhibit axial symmetry. In this appendix barodesy for clay is presented for axially symmetric conditions, i.e. $D_2 = D_3$ and $T_2 = T_3$. The axial stress rate $\dot{T}_1$ and the radial stress rate $\dot{T}_2$ are then given by the following scalar functions:

$$
\dot{T}_1 = c_3 \left(T_1^2 + 2T_2^2 \right)^{c_4/2} \left(f \frac{R_1}{\sqrt{R_1^2 + 2R_2^2}} + g \frac{T_1}{\sqrt{T_1^2 + 2T_2^2}} \right) \sqrt{D_1^2 + 2D_2^2}
$$

$$
\dot{T}_2 = c_3 \left(T_1^2 + 2T_2^2 \right)^{c_4/2} \left(f \frac{R_2}{\sqrt{R_1^2 + 2R_2^2}} + g \frac{T_2}{\sqrt{T_1^2 + 2T_2^2}} \right) \sqrt{D_1^2 + 2D_2^2}
$$

The directions of proportional stress paths in axial and in radial directions are the scalars R_1 and R_2:

$$
R_1 = -\exp\left(\alpha \frac{D_1}{\sqrt{D_1^2 + 2D_2^2}} \right), \quad R_2 = -\exp\left(\alpha \frac{D_2}{\sqrt{D_1^2 + 2D_2^2}} \right)
$$

$$
\text{with} \quad \alpha = \frac{\ln K}{\sqrt{3/2 - \delta^2/2}}, \quad K = 1 - \frac{1}{1 + c_1(m - c_2)}, \quad m = \frac{-3\delta}{\sqrt{6 - 2\delta^2}}
$$

The dilatancy δ equals $\frac{D_1 + 2D_2}{\sqrt{D_1^2 + 2D_2^2}}$. The scalar quantities f and g are:

$$
f = c_6 \cdot \beta \cdot \delta - \frac{1}{2}
$$

$$
g = (1 - c_6) \cdot \beta \cdot \delta + \left(\frac{1 + e}{1 + e_c} \right)^{c_5} - \frac{1}{2}
$$

$$
\text{with} \quad e_c = \exp\left[N - \lambda^* \ln \left(-\frac{2(T_1 + 2T_2)}{3\sigma^*} \right) \right] - 1
$$

$$
\beta = -\frac{1}{c_3 \Lambda} + \frac{1}{\sqrt{3}} 2^{c_5 \lambda^*} - \frac{1}{\sqrt{3}}
$$

$$
\Lambda = -\frac{\lambda^* - \kappa^*}{2\sqrt{3}} \delta + \frac{\lambda^* - \kappa^*}{2}
$$

C.1

The material constants c_1 - c_6 are determined according to Table 5.2. N, λ^* and κ^* are material parameters. The reference pressure σ^* equals $1\,\text{kPa}$. The functions f and g as well as the tensorial equation are described in detail in Chapters 4 and 5.

Appendix D

Source code

In this appendix the MATLAB source code of barodesy is presented. Listing D.1 shows how the material constants based on Critical State Soil Mechanics parameters are determined, see also Table 4.1 on page 52. In lines 5 - 10 the Critical State Soil Mechanics parameters (φ_c, N, λ^*, κ^* and G at the respective stress level p) are filled in. Based on those parameters, the constants c_1 - c_6 are determined in lines 13 - 23. Note that c_4 equals 1 for clay and is therefore omitted here.

```
1  function setC()
2  global  phi_c N lambda kappa c_1 c_2 c_3 c_5 c_6
3
4  % Input: Fill in Critical State Soil Mechanics parameters:
5  phi_c  = 22.6;   % in degrees
6  N      = 1.375;
7  lambda = 0.11;
8  kappa  = 0.016;
9  G      = 11000;  % initial shear stiffness G (kPa) ...
10 p      = 110;    % ... at the respective stress level p (kPa)
11
12 % Output: Calculate constants c_1 - c_6 by CSSM paramters
13 K_c   = (1-sind(phi_c))/(1+sind(phi_c));
14 K_0   = 1-sind(phi_c);
15 c_2   = (3*sqrt(K_c*(1-K_c)*K_0*(1-K_0))+...
16          3*K_c*(1-K_0))/(2*(K_c-K_0));
17 c_1   = -K_c/(c_2^2*(K_c-1));
18 c_3_1 = 12*G*(1+2*K_c^2)^(1/2)/(sqrt(6)*(sqrt(3)*p)*(K_c-1));
19 c_3_2 = (-sqrt(3)/lambda+sqrt(3)/kappa)/(2^(1/K_c*lambda)+...
20          (2*1/100)^(lambda*1/K_c)-2);
21 c_3   = max(c_3_1,c_3_2);
22 c_5   = 1/K_c;
23 c_6   = 1/(2*(-sqrt(3)/kappa/c_3-(1-2^(lambda*c_5))));
```

Listing D.1: Determination of material parameter, shown for London clay

Listings D.2 and D.3 include the constitutive model.

```
1  function [Tp,ep] = getobjtr_baroclay(T,e,D)
2  global  N lambda kappa c_3 c_5 c_6
3
4  % abbreviations:
5  nrmD  = norm(D,'fro');  D0 = D/nrmD;  trD0 = trace(D0);
```

```
6  nrmT    = norm(T,'fro'); T0 = T/nrmT; p      = -trace(T)/3;
7
8  % Barodesy:
9  ec      = exp(N-lambda*log(2*p))-1; % critical  void ratio
10 Lambda  = -(lambda-kappa)/2/sqrt(3)*trD0+(lambda+kappa)/2;
11 beta    = -1/Lambda/c_3+2^(c_5*lambda)/sqrt(3)-1/sqrt(3);
12 f       = c_6*beta*trD0 -1/2;
13 g       = ((1+e)/(1+ec))^c_5+(1-c_6)*beta*trD0 -1/2;
14 R       = get_R(D0); R0 = R/norm(R,'fro');
15 Tp      = c_3 * nrmT * (f * R0 + g * T0) * nrmD;
16 ep      = (1 + e) * trace(D); % change of void ratio
```

Listing D.2: Source code for barodesy for clay

```
1  % R - function: calculating  directions  of proportional  stress  paths
2  function R = get_R(D0)
3  global    c_1  c_2
4
5  zero = 10^-10;
6
7  % invariants of D0:
8  I_1     = trace(D0);                      % 1st invariant of D0
9  % zero is added -> the root is calculated from a positive value:
10 D_dev = (3/2-I_1^2/2+zero)^(1/2);    % distortional  stretching
11 m       = -I_1/(2/3*D_dev);          % strain  increment  ratio
12
13 % R - function:
14 K       = -1/(1 + c_1*(m - c_2)^2)+1;
15 alpha = log(K+zero)/(D_dev+zero);    % zero is added
16 R       = -expm(alpha*D0);
```

Listing D.3: Source code of the **R**-function: Determination of PσPs

In Listing D.3 the directions of proportional stress paths (**R**) are calculated. In lines 8 - 11 invariants of $\mathbf{D}^0$ are determined. Note that in line 10 a small number (10^{-10}, see line 5) is added in order to assure that the root is calculated from a positive value. Thus numerical errors are kept small. The same applies for the determination of α in line 15. Note that in line 16 the matrix exponential (expm) is computed.

As an example, stress paths obtained by proportional strain paths are simulated (Listing D.4) and it is shown how the subroutines according to Listings D.1, D.2 and D.3 are integrated. In line 4 the function setC (Listing D.1) is called and the material constants are set. In line 27 the constitutive model (getobjtr_baroclay, see Listing D.2) is called. In line 14 in getobjtr_baroclay the subroutine get_R (Listing D.3) is run. Figure D.1 shows the results obtained by Listing D.4.

```
1  % stress path obtained by proportional strain paths
2  close all; clear all;
3  global N lambda
4  setC;                              % set material constants
5  T_0 = -200*eye(3,3);              % initial stress state
6  p   = -1/3*(T_0(1,1)+T_0(2,2)+T_0(3,3));
7  e_0 = exp(N-lambda*log(p)) -1.15; % initial void ratio
8  D   = zeros(3,3);                 % stretching
9  Dt  = 5e-4;                       % time step
10
11 % calculate stress paths
12 j=1;
13 for alpha=0:22.5:359              % in degrees
14   e       = e_0;
15   D(1,1) = -sind(alpha);
16   D(2,2) = -cosd(alpha)/sqrt(2);
17   D(3,3) = D(2,2);
18   T = T_0;
19   sigma_plot(j).T1 = -T(1,1);     % defining plot variables
20   sigma_plot(j).T2 = -T(2,2);
21   D_plot(j,:) = -[D(1,1) D(2,2) D(3,3)];
22 % the  "while loop" is left, when abs(T_11) or abs(T_22) are
23 % either smaller 0.01kPa or larger 500 kPa.
24 % the maximum number of integration steps is 1000
25   while(T(1,1) < -.01 && T(2,2) < -.01 && T(1,1) > -500...
26         && T(2,2) > -500) && length(sigma_plot(j).T1) < 1000
27         [Tp,ep] = getobjtr_baroclay(T,e,D); % stress rate
28              T = T + Tp*Dt;                 % actual stress
29              e = e + ep*Dt;                 % actual void ratio
30              sigma_plot(j).T1(end+1) = -T(1,1);
31              sigma_plot(j).T2(end+1) = -T(2,2);
32   end
33 j=j+1;
34 end
35 % plot stress paths
36 figure(1); subplot(1,2,1)
37 for i=1:j-1
38   plot([0 D_plot(i,2)]*sqrt(2),[0 D_plot(i,1)]); hold on
39 end
40 axis square; box off
41 %
42 subplot(1,2,2)
43 for i=1:j-1
44   plot(sigma_plot(i).T2*sqrt(2),sigma_plot(i).T1); hold on
45 end
46 axis([0 500 0 500]); axis square; box off
```

Listing D.4: Source code to calculate stress path obtained by constant stretching, see Figure D.1.

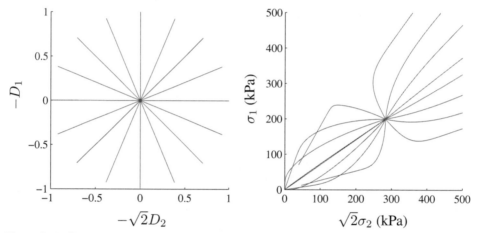

Figure D.1: Stress paths obtained by proportional strain paths, calculated and plotted according to Listing D.4.

Appendix E

Basic clay features

In this appendix basic clay features of different clays (Dresden clay, Weald clay, Dortmund clay, Kaolin clay and Fujinomori clay) are shown. The material constants are in Table 5.3 on page 74. The initial stress state for Figures (a) and (d) is $\mathbf{T}_{\text{ini}} = -50 \cdot \mathbf{1}$ kPa. The initial void ratio in Figures (a), (c) and (d) is $e_{\text{ini}} = \exp\left(N - \lambda^* \ln(2p/\sigma^*)\right) - 1$. Figures (a) show the state boundary surfaces for different clays. In Figures (b) Chu & Lo's stress-dilatancy relation is compared with predictions by barodesy.

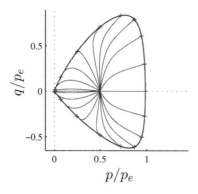

(a) State boundary surface

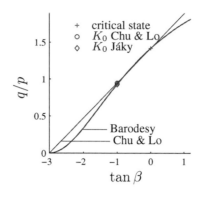

(b) Chu & Lo's predictions

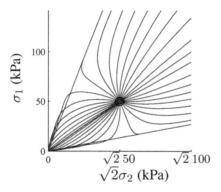

(c) Starting from $\mathbf{T} \neq \mathbf{0}$ and applying $P\varepsilon Ps$

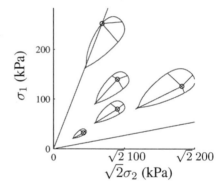

(d) Response envelopes

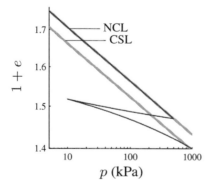

(e) Loading, unloading and reloading under isotropic compression in the $\ln p$ - $\ln(1+e)$ plot

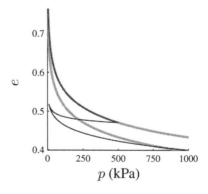

(f) Loading, unloading and reloading under isotropic compression in the p - e plot

Figure E.1: Dresden Clay, simulated with barodesy

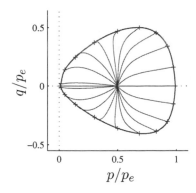

(a) State boundary surface

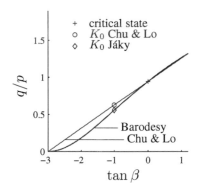

(b) Chu & Lo's predictions

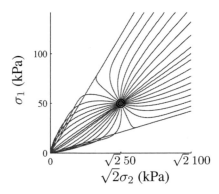

(c) Starting from $\mathbf{T} \neq \mathbf{0}$ and applying $P\varepsilon Ps$

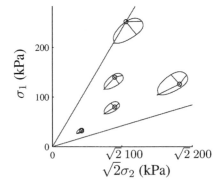

(d) Response envelopes

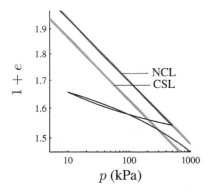

(e) Loading, unloading and reloading under isotropic compression in the $\ln p$ - $\ln(1+e)$ plot

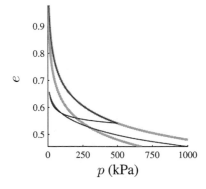

(f) Loading, unloading and reloading under isotropic compression in the p - e plot

Figure E.2: Weald Clay, simulated with barodesy

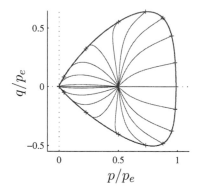

(a) State boundary surface

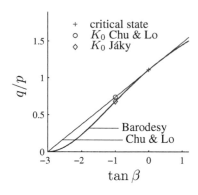

(b) Chu & Lo's predictions

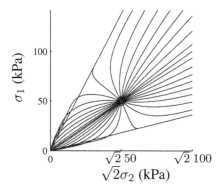

(c) Starting from $\mathbf{T} \neq \mathbf{0}$ and applying PεPs

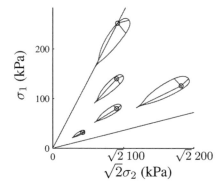

(d) Response envelopes

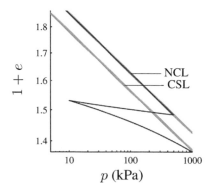

(e) Loading, unloading and reloading under isotropic compression in the $\ln p$ - $\ln(1+e)$ plot

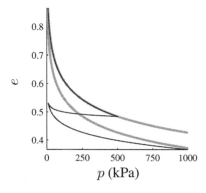

(f) Loading, unloading and reloading under isotropic compression in the p - e plot

Figure E.3: Dortmund Clay, simulated with barodesy

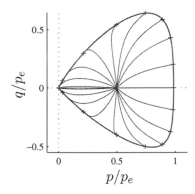

(a) State boundary surface

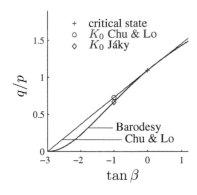

(b) Chu & Lo's predictions

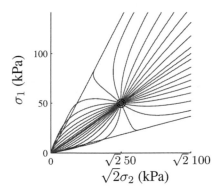

(c) Starting from $\mathbf{T} \neq \mathbf{0}$ and applying $P\varepsilon Ps$

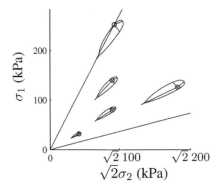

(d) Response envelopes

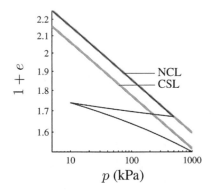

(e) Loading, unloading and reloading under isotropic compression in the $\ln p$ - $\ln(1 + e)$ plot

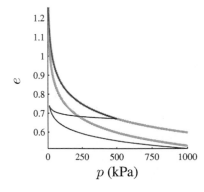

(f) Loading, unloading and reloading under isotropic compression in the p - e plot

Figure E.4: Kaolin Clay, simulated with barodesy

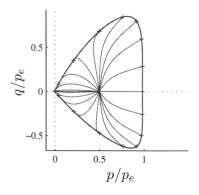

(a) State boundary surface

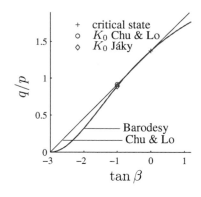

(b) Chu & Lo's predictions

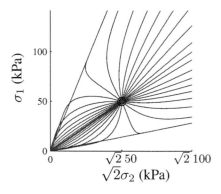

(c) Starting from $\mathbf{T} \neq \mathbf{0}$ and applying $P\varepsilon Ps$

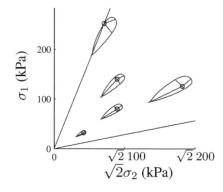

(d) Response envelopes

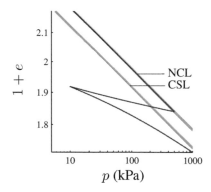

(e) Loading, unloading and reloading under isotropic compression in the $\ln p$ - $\ln(1+e)$ plot

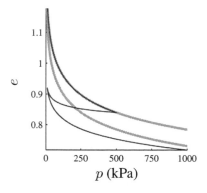

(f) Loading, unloading and reloading under isotropic compression in the p - e plot

Figure E.5: Fujinomori Clay, simulated with barodesy